Tucholsky Wagner Zola Scott Sydow Freud Schlegel
Turgenev Wallace Fonatne
Twain Walther von der Vogelweide Fouqué Friedrich II. von Preußen
Weber Freiligrath Frey
Fechner Fichte Weiße Rose von Fallersleben Kant Ernst Frommel
Richthofen
Engels Fielding Hölderlin
Fehrs Faber Flaubert Eichendorff Tacitus Dumas
Maximilian I. von Habsburg Fock Eliasberg Ebner Eschenbach
Feuerbach Zweig
Ewald Eliot Vergil
Goethe Elisabeth von Österreich London
Mendelssohn Balzac Shakespeare Dostojewski Ganghofer
Trackl Lichtenberg Rathenau Doyle Gjellerup
Mommsen Stevenson Tolstoi Hambruch Droste-Hülshoff
Thoma Lenz Hanrieder
Dach Verne von Arnim Hägele Hauff Humboldt
Reuter Rousseau Hagen Hauptmann Gautier
Karrillon Garschin Defoe Baudelaire
Damaschke Hebbel
Descartes Hegel Kussmaul Herder
Wolfram von Eschenbach Dickens Schopenhauer
Bronner Darwin Melville Grimm Jerome Rilke George
Campe Horváth Aristoteles Bebel Proust
Bismarck Vigny Voltaire Federer Herodot
Gengenbach Barlach Heine
Storm Casanova Tersteegen Grillparzer Georgy
Chamberlain Lessing Langbein Gilm Gryphius
Brentano Lafontaine
Strachwitz Claudius Schiller Kralik Iffland Sokrates
Katharina II. von Rußland Bellamy Schilling
Gerstäcker Raabe Gibbon Tschechow
Löns Hesse Hoffmann Gogol Wilde Vulpius
Luther Heym Hofmannsthal Gleim
Roth Klee Hölty Morgenstern Goedicke
Luxemburg Heyse Klopstock Kleist
La Roche Puschkin Homer Mörike
Machiavelli Horaz Musil
Navarra Aurel Musset Kierkegaard Kraft Kraus
Lamprecht Kind Moltke
Nestroy Marie de France Kirchhoff Hugo
Laotse Ipsen Liebknecht
Nietzsche Nansen Ringelnatz
Marx Lassalle Gorki Klett
von Ossietzky Leibniz
May vom Stein Lawrence Irving
Petalozzi Knigge
Platon Michelangelo Kafka
Sachs Pückler Kock
Poe Liebermann Korolenko
de Sade Praetorius Mistral Zetkin

The publishing house tredition has created the series **TREDITION CLASSICS**. It contains classical literature works from over two thousand years. Most of these titles have been out of print and off the bookstore shelves for decades.

The book series is intended to preserve the cultural legacy and to promote the timeless works of classical literature. As a reader of a **TREDITION CLASSICS** book, the reader supports the mission to save many of the amazing works of world literature from oblivion.

The symbol of **TREDITION CLASSICS** is Johannes Gutenberg (1400 – 1468), the inventor of movable type printing.

With the series, tredition intends to make thousands of international literature classics available in printed format again – worldwide.

All books are available at book retailers worldwide in paperback and in hardcover. For more information please visit: www.tredition.com

tredition was established in 2006 by Sandra Latusseck and Soenke Schulz. Based in Hamburg, Germany, tredition offers publishing solutions to authors and publishing houses, combined with worldwide distribution of printed and digital book content. tredition is uniquely positioned to enable authors and publishing houses to create books on their own terms and without conventional manufacturing risks.

For more information please visit: www.tredition.com

A Handbook of Some South Indian Grasses

K. Rangachari

Imprint

This book is part of the TREDITION CLASSICS series.

Author: K. Rangachari
Cover design: toepferschumann, Berlin (Germany)

Publisher: tredition GmbH, Hamburg (Germany)
ISBN: 978-3-8491-5568-1

www.tredition.com
www.tredition.de

Copyright:
The content of this book is sourced from the public domain.

The intention of the TREDITION CLASSICS series is to make world literature in the public domain available in printed format. Literary enthusiasts and organizations worldwide have scanned and digitally edited the original texts. tredition has subsequently formatted and redesigned the content into a modern reading layout. Therefore, we cannot guarantee the exact reproduction of the original format of a particular historic edition. Please also note that no modifications have been made to the spelling, therefore it may differ from the orthography used today.

PREFACE

This book is intended to serve as a guide to the study of grasses of the plains of South India. For the past few years I have been receiving grasses for identification, almost every week, from the officers of the Agricultural and Forest Departments and others interested in grasses. The requirements of these men and the absence of a suitable book induced me to write this book.

I have included in this book about one hundred grasses of wide distribution in the plains of South India. Many of them occur also in other parts of India. The rarer grasses of the plains and those growing on the hills are omitted, with a view to deal with them separately.

The value of grasses can be realized from the fact that man can supply all his needs from them alone, and their importance in agriculture is very great, as the welfare of the cattle is dependent upon grasses. Farmers, as a rule, take no interest in them, although profitable agriculture is impossible without grasses. Very few of them can give the names of at least half a dozen grasses growing on their land. They neglect grasses, because they are common and are found everywhere. They cannot discriminate between them. To a farmer "grass is grass" and that is all he cares to trouble himself about. About grasses Robinson writes "Grass is King. It rules and governs the world. It is the very foundation of all commerce: without it the earth would be a barren waste, and cotton, gold, and commerce all dead."

In the early days when the population was very much limited and when land not brought under cultivation was extensive plenty of green grasses was upon it and pastures were numerous. So the farmer paid no attention to the grasses, and it did not matter much. But now, population has increased, unoccupied land has decreased very much and the cattle have increased in number. Consequently he has to pay more attention to grasses.

On account of the scarcity of fodder, people interested in agriculture and cattle rearing have very often imported [Pg iv] foreign grasses and fodder plants into this country, but so far no one has succeeded in establishing any one of them on any large scale. Usually a great amount of labour and much money is spent in these attempts. If the same amount of attention is bestowed on indigenous grasses, better results can be obtained with less labour and money. There are many indigenous grasses that will yield plenty of stuff, if they are given a chance to grow. The present deterioration of grasses is mainly due to overgrazing and trampling by men and cattle.

To prove the beneficial effects which result from preventing overgrazing and trampling, Mr. G. R. Hilson, Deputy Director of Agriculture (now Cotton Expert), selected some portion of the waste land in the neighbourhood of the Farm at Hagari and closed it for men and cattle. As a result of this measure, in two years, a number of grasses and other plants were found growing on the enclosed area very well, and all of them seeded well. Of course the unenclosed areas were bare as usual.

In the preparation of this book I received considerable help from M.R.Ry. C. Tadulinga Mudaliyar Avargal, F.L.S., Assistant Lecturing and Systematic Botanist, in the description of species and I am indebted to M.R.Ry. P.S. Jivanna Rao, M.A., Teaching Assistant, for assistance in proofreading.

I have to express my deep obligation to Mr. G. A. D. Stuart, I.C.S., Director of Agriculture, for encouragement to undertake this work and to the Madras Government for ordering its publication.

For the excellence in the get up of the book I am indebted to Mr. F. L. Gilbert, Superintendent, Government Press.

K. RANGACHARI.

Agricultural College,
Lawley Road, Coimbatore,
2nd June 1921.

[Pg v]

CONTENTS

[Pg 1]

A HANDBOOK OF SOME COMMON SOUTH INDIAN GRASSES.

CHAPTER I.
INTRODUCTION.

Grasses occupy wide tracts of land and they are evenly distributed in all parts of the world. They occur in every soil, in all kinds of situations and under all climatic conditions. In certain places grasses form a leading feature of the flora. As grasses do not like shade, they are not usually abundant within the forests either as regards the number of individuals, or of species. But in open places they do very well and sometimes whole tracts become grass-lands. Then a very great portion of the actual vegetation would consist of grasses.

On account of their almost universal distribution and their great economic value grasses are of great importance to man. And yet very few people appreciate the worth of grasses. Although several families of plants supply the wants of man, the grass family exceeds all the others in the amount and the value of its products. The grasses growing in pasture land and the cereals grown all over the world are of more value to man and his domestic animals than all the other plants taken together.

To the popular mind grasses are only herbaceous plants with narrow leaves such as the hariali, ginger grass and the kolakattai grass. But in the grass family or Gramineæ the cereals, sugarcane and bamboos are also included.

Grasses are rather interesting in that they are usually successful in occupying large tracts of land to the exclusion of other plants. If we take into consideration the number of individuals of any species of grass, they will be found to out-number those of any species of any other family. Even as regards the number of species this family ranks fifth, the first four places being occupied respectively by Compositæ, Leguminosæ, Orchideæ and Rubiaceæ.

As grasses form an exceedingly natural family it is very difficult for beginners to readily distinguish them from one another.

The leaves and branches of grasses are very much alike and the flowers are so small that they are liable to be passed by unnoticed. The recognition of even our common grasses is quite a task for a botanist.

To understand the general structure of grasses and to become familiar with them it is necessary to study closely some common grasses. We shall begin our study by selecting as a type one of the species of the genus Panicum.

Panicum javanicum is an annual herb with stems radiating in all directions from a centre. The plant is fixed to the soil by a tuft of [Pg 2] fibrous roots all springing from the bases of the stems. In addition to this crown of fibrous roots, there may be roots at the nodes of some of the prostrate branches. The stems and branches are short at first, and leaves arise on them one after the other in rapid succession. After the appearance of a fair number of leaves the stem elongates gradually and it finally ends in an inflorescence.

Fig. 1. — Panicum javanicum. (Full plant.)

The stem consists of **nodes** and **internodes**. The internodes are cylindrical and somewhat flattened on the side towards the axillary bud. When young they are completely covered by the leaves and the older ones have only their lower portions covered by the leaf-sheaths. Usually they complete their growth in length very soon, but the lower portion of the internode, just above the node and enclosed by the sheath, retains its power of growth for some time.

The leaf consists of the two parts, the **leaf-sheath** and the **leaf-blade**. At the junction of these two parts there is a very thin [Pg 3] narrow membrane with fine hairs on its free margin. This is called the **ligule**. (See fig. 2.)

The leaf-sheath is attached at its base to the node and it is slightly swollen just above the place of insertion. It covers the internode, one margin being inside and the other outside. The surface of the sheath is sparsely covered with long hairs springing from small tubercles. The outer margin of the sheath bears fine hairs all along its length. (See fig. 2.)

The leaf-blade is broadly lanceolate, with a tip finely drawn out. Its base is rounded and the margin wavy, especially so towards the base. On the margin towards the base long hairs are seen, and some of these arise from small tubercles. The margin has a hyaline border which is very minutely serrate. There is a distinct midrib and, on holding the leaf against the light, four or five small veins come in to view. In the spaces between these veins lie many fine veins. All the veins run parallel from the base to the apex. At the base of the blade the veins get into the leaf-sheath and therefore the sheath becomes striated. Just above the ligule and at the base of the leaf-blade there is a colourless narrow zone. This is called the **collar**.

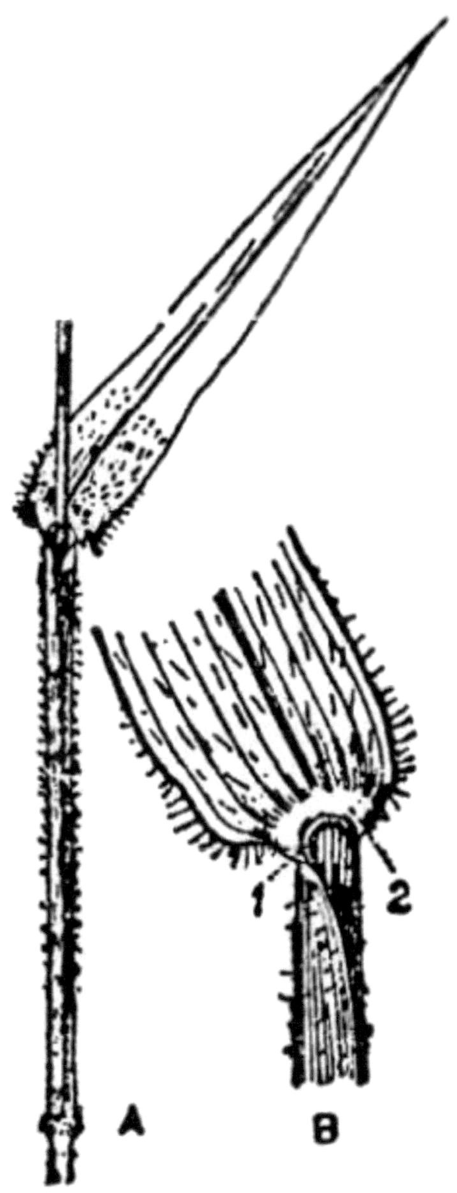
A
B
1
2

Fig. 2.—Leaf of Panicum javanicum.
A. Full leaf; B. a portion of the leaf showing 1. the ligule and 2. the collar.

As already stated the inflorescences appear at the free ends of branches. Every branch sooner or later terminates in an inflorescence which is a compound raceme. There are usually five or six racemes in the inflorescence. Each raceme has an axis, called the **rachis**, which bears unilaterally two rows of bud-like bodies. These bud-like bodies are the units of the inflorescence and they are called **spikelets**. (See fig. 3.)

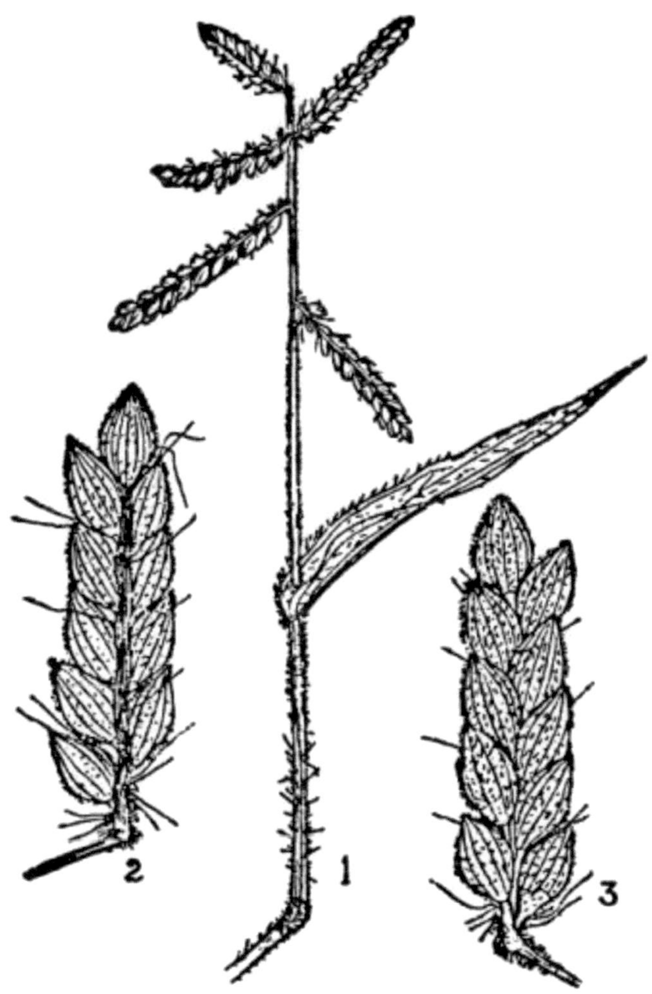

Fig. 3. — The inflorescence of Panicum javanicum.
1. Inflorescence; 2 and 3. the front and the back view of a raceme.

[Pg 4]

The spikelets are softly hairy and are shortly stalked. The pedicels
of spikelets are hairy and sometimes one or two long hairs are also

found on them. Each of these spikelets consists of four green membranous structures called **glumes**. The first two glumes are unequal, the first being very small. The second and the third glumes are broadly ovate-oblong with acute tips. Both are of the same height and texture, but the second is 7-nerved and the third 5-nerved. The fourth glume is membranous when young, but later on it becomes thick, coriaceous and rugose at the surface. Just opposite to the fourth glume there is a flat structure with two nerves, similar to the glume in texture. This is called the **palea**. The fourth glume and its palea adhere together by their margins. Inside the fourth glume and between it and the palea there are three stamens and an ovary with two styles ending in feathery stigmas. Just in front of the ovary and outside the stamens two very small scale-like bodies are found. These are the **lodicules**. They are fleshy and well developed in flowers that are about to open. In the spikelet there is only one full flower. The third glume contains no flower in it, but occasionally there may be in its axil three stamens. The first two glumes are always empty and so they are called empty glumes. (See fig. 4.) In mature spikelets the grain which is free is enclosed by the fourth glume and its palea.

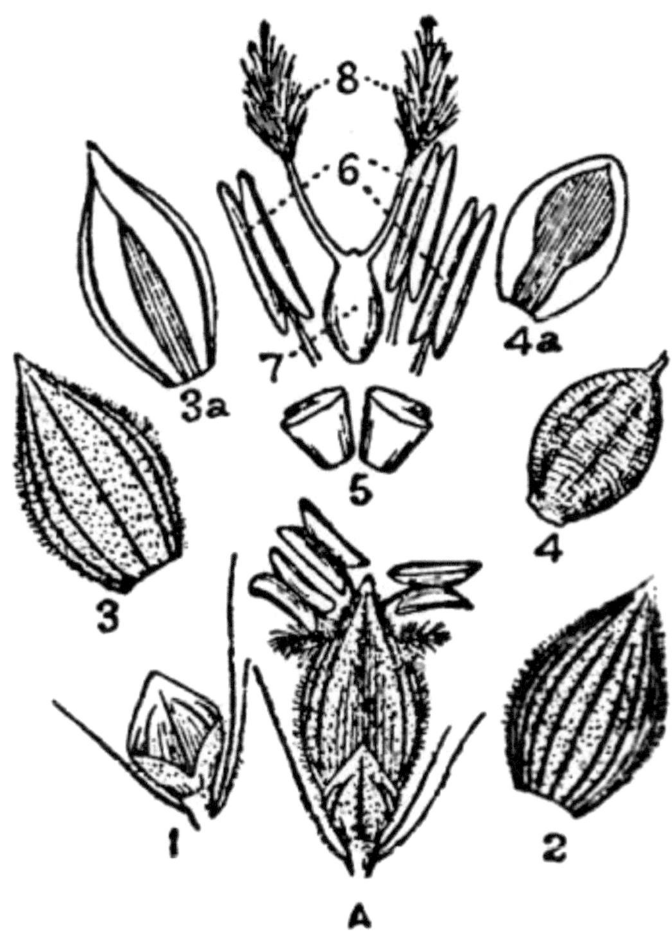

Fig. 4.—Parts of the spikelets of Panicum javanicum.
A. A spikelet; 1, 2, 3 and 4. the first, second, third and the fourth
glume, respectively; 3a. palea of the third glume; 4a. palea of the
fourth glume; 5. lodicules; 6. stamens; 7. ovary; 8. stigmas.

[Pg 5]

CHAPTER II.
THE VEGETATIVE ORGANS.

Grasses vary very much in their habit. Some grasses grow erect forming tufts and others form cushions with the branches creeping along the ground. (See figs. 5 and 6.) We usually find all intermediate stages from the erect to the prostrate habit. Underground stems such as stolons and rhizomes occur in some grasses. Grasses of one particular species generally retain the same habit but this does not always hold good. For example *Trugus racemosus* grows with all its branches quite prostrate in a poor, dry, open soil. If, on the other hand, this happens to grow in rich soils, or amidst other plants or grasses, it assumes an erect, somewhat tufted habit. *Andropogon contortus* and *Andropogon pertusus* are other grasses with a tendency for variation in habit. Plants that are usually small often attain large dimensions under favourable conditions of growth. Ordinarily the grass *Panicum javanicum* grows only to 1 or 2 feet. (See fig. 1.) The same plant in a good rich soil grew to about 6 feet in four months. (See fig. 7.)

Fig. 5.—Eleusine ægyptiaca.

Some grasses are annual while others are perennial. It is often difficult to determine whether a certain grass is annual or perennial. But by examining the shoot-system this can be ascertained easily. In an annual all the stems and branches usually end in inflorescences and they will all be of the same year. If, on the other hand, both young leafy branches and old branches ending in inflorescences are found mixed, it must be a perennial grass. The presence of the remains of old leaves, underground stolons and rhizomes is also evidence showing the perennial character of the plant.

Grasses are eminently adapted to occupy completely large areas of land. They are also capable of very rapid extension over large areas, on account of the production of stolons, rhizomes and the formation of adventitious roots.

[Pg 6]

The root-system. — The root-system of grasses is very striking in its character. In most grasses, especially in erect ones, several roots all of about the same diameter arise in a dense tuft from nearly the same level and from the lower-most nodes of the stems. The roots are all thin and fibrous in the vast majority of these plants, and they are tough and wiry only in a few cases such as in the case of the roots of *Pennisetum cenchroides, P. Alopecuros, Ischæmum pilosum* and *Andropogon Schœnanthus.*

On a close examination it will become evident that all the roots of a grass plant are adventitious. Inasmuch as the growth of the primary root is soon overtaken by other roots growing from the stem, all the roots happen to be of the same size. Roots arise from the nodes just above the insertion of the leaf, and they grow piercing the leaf-sheath.

Fig. 6. — Panicum Crus-galli.

Grasses in which stolons and prostrate branches occur have, in addition to the usual radiating crown of roots at the base, aerial roots growing out of the upper nodes of the branches and fixing them to the soil. Such roots become supporting or prop roots and are particularly conspicuous in several stout tall grasses such as

Andropogon Sorghum, Zea Mays and *Pennisetum typhoideum.* (See figs. 8 and 9.)

All the roots bear branch-roots which originate from the inner portion of the mother roots in the usual manner. The character and the extent of the development of the root-system is to a large extent dependent upon the nature of the soil and its moisture content. In light dry soils roots remain generally stunted and in [Pg 7] well drained rich soils they attain their maximum development. In clay-ey soils roots penetrate only to short distances. When the soil is rich and sandy roots go deeper and extend in all directions. The root-systems of most grasses are superficial and so are best adapted for surface-feeding.

Fig. 7. — Panicum javanicum.

The shoot-system.—The shoot-system varies with the duration of the life of the plant. In annual grasses stems are in most cases [Pg 8] erect and even if they are not entirely so they become erect at the time of flowering. They are attached to the soil by a tuft of fibrous roots arising from the base of the stems. But in perennials in addition to erect branches, creeping branches, stolons and rhizomes may occur.

Fig. 8.—Prop roots of Andropogon Sorghum.

Fig. 9. — Aerial roots of Ischæmum ciliare.

The stem is either cylindrical or compressed and consists of nodes and internodes. In most grasses the internodes are usually hollow, the cavity being lined by the remains of the original pith cells. However, there are also grasses in which the stems remain solid throughout. In many grasses the basal portions of stems are more leafy and the internodes are short, but in the upper portions the internodes become longer separating the leaves one from the other.

In young shoots the leaves grow much faster than the internodes and consequently internodes remain small, and leaves become very conspicuous. The youngest portions of the shoots are by this means always well protected by the surrounding leaf-sheaths. As soon as leaves have grown fully, the internodes begin to elongate rapidly separating the leaves. At first growth in length takes place throughout its length in the internode and when it gets older this elongation ceases. But, however, the lower portion of the internode close to the node and which is enclosed by the leaf-sheath retains its power of growth for a considerable time.

Branches arise from the axils of leaves and when a considerable number of the axillary buds, especially from the lower nodes, develop into branches the plant becomes tufted in habit. In most grasses branches grow upwards through the sheath and emerge at its mouth as aerial branches. Such branches are called **intravaginal** branches or stems. But in some grasses axillary buds, [Pg 9] instead of growing straight up through the sheath, pierce the leaf-sheath, come out and then they grow out as branches. This may be seen in the underground stolons of *Panicum repens* and in the ordinary aerial branches of *Arundo Donax*. Branches that pierce through the sheaths are called **extravaginal** branches. (See fig. 10.)

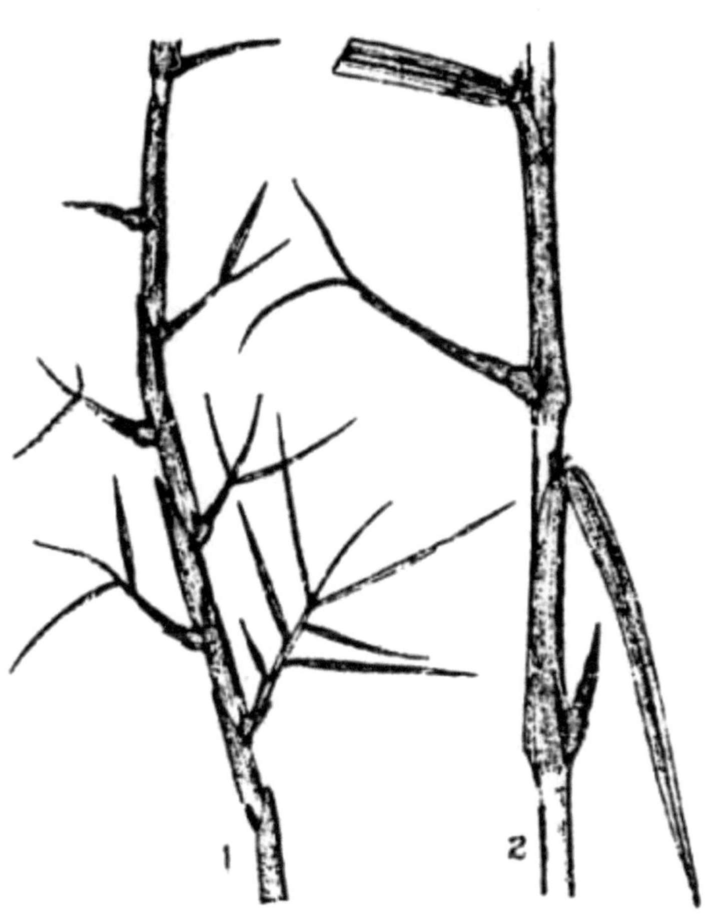

Fig. 10. — Extravaginal shoots of 1. Panicum repens and 2. Arundo Donax.

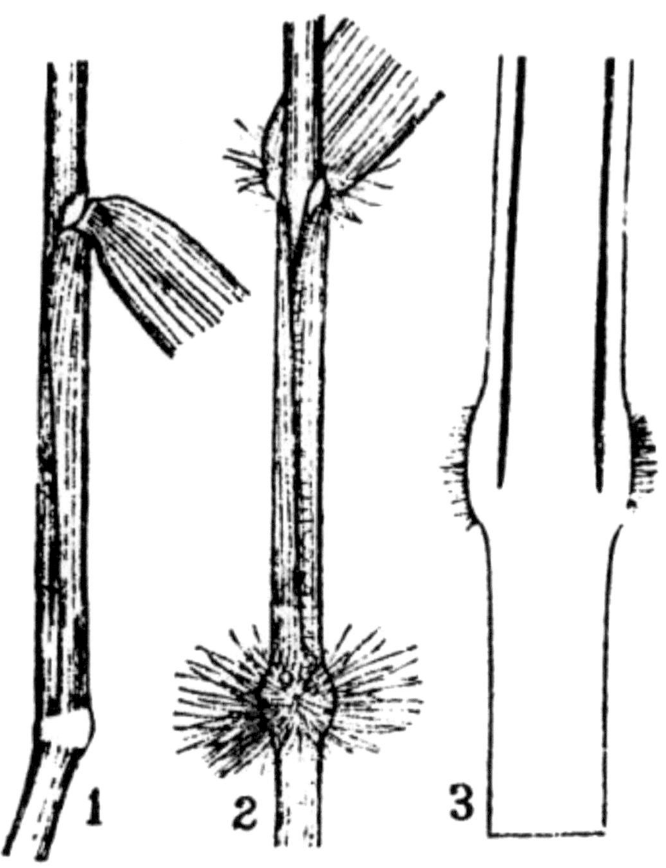

Fig. 11.—Nodes.
1. Glabrous node; 2. bearded node; 3. node cut longitudinally.

The nodes are in most cases very conspicuous and they are often found swollen. However, it must be remembered that the enlargement at the node is not due to the increase in size of the actual node, but due to growth in thickness of the base of the leaf-sheath. (See fig. 11-3.) Nodes may be pale or coloured, glabrous, hairy or bearded with long hairs. When the stem is erect the nodes are short and of uniform size all round. But, if the stem is bent down or tipped over by accident, the nodes begin to grow longer on the lower side until a curvature sufficient to bring the stem to the erect position is formed and then it ceases to grow.

As already noted some perennial grasses have creeping stems and stolons, while others may have rhizomes. The grass *Cynodon dactylon* develops several underground stolons which are covered with white scale leaves and whose terminal buds are hard and sharp so that they may be able to make their way through the soil. The rhizomes when continuous and elongated are usually sympodia formed by the lower portions of the aerial shoots. The aerial shoot comes into the air and its lower portion is continued by a branch arising from a lower leaf axil beneath the soil.

The leaf.—Leaves are two-ranked and alternate, and very often they become crowded at the lower portions of the shoots so as to form basal tufts, though they are farther apart in the upper [Pg 10] portions of these shoots. Three distinct kinds of leaves are met with in grasses. First, we have the fully formed foliage leaves so characteristic of grasses. These are most conspicuous and are formed in large numbers.

The other two kinds of leaves are neither so conspicuous nor so numerous as the foliage leaves. At the base of shoots occur abortive leaves which are really rudimentary sheaths. These are called **scales**. The third kind of leaf is a modified structure called the **prophyll** or **prophyllum**. (See fig. 12.) It is the first leaf occurring in every branch on the side next to the main shoot and it is a two-keeled membranous structure resembling somewhat the palea found in the spikelets of grasses. The portion of the prophyll between the keels is concave due to the pressure of the main stem, while the sides beyond the keels bend forward clasping the stem.

32

Fig. 12. — Prophylla.
A. A branch with its prophyllum; B. prophyllum; C. section of the prophyllum.

The ordinary foliage leaves of grasses consist of the two parts, the flat expanded upper portion called the **blade** and the lower part called the **sheath** that encircles the stem above the node from which it arises. The leaf-sheaths usually fit close to the stem, but they may also be loose or even inflated. Though the leaf-sheath surrounds the internode like a tube, it is not a closed tube. It is really a flat structure rolled firmly round the stem with one edge overlapping the other. In most cases it is cylindrical and it may be compressed in a few cases. Occasionally it may have a prominent ridge or keel down its back. The sheath may be glabrous or hairy, smooth or striate externally, and the outer margin is often ciliate. In a few grasses the sheaths become coloured especially below or on the side exposed to the sun.

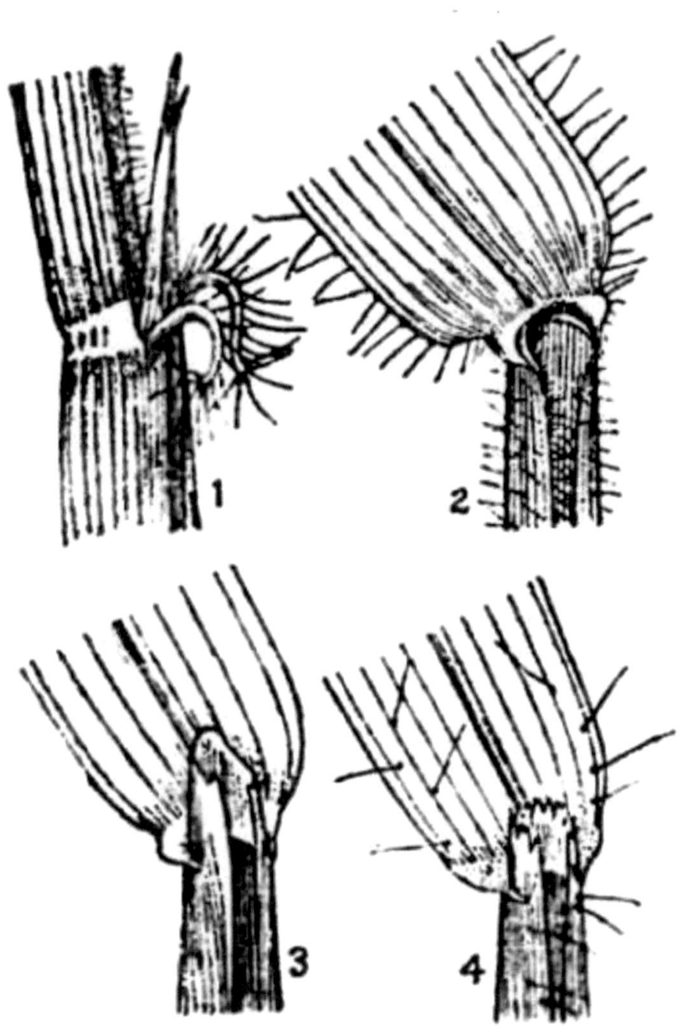

Fig. 13.—Ligules of 1. Oryza sativa; 2. Panicum javanicum; 3. Andropogon Schœnanthus; 4. A. contortus.

[Pg 11]

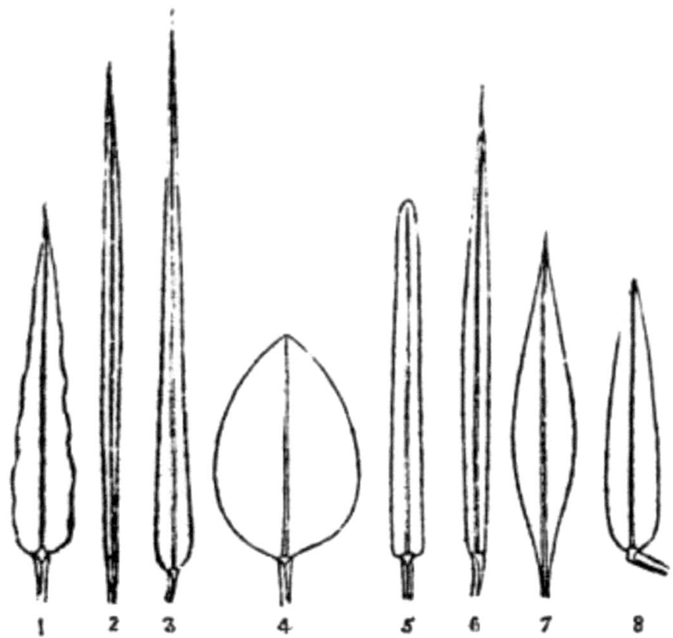

Fig. 14.—Shapes of leaf-blades.
1, 7 and 8. Lanceolate; 3 and 6. lanceolate-linear; 2 and 5. linear; and 4. ovate.

The **ligule** is a structure peculiar to grasses and it varies very much. In some grasses it is a distinct membrane narrow or broad, with an even, truncate or erose margin, or finely ciliate. Very often it is only a line or fringe of hairs, whilst in some it may be entirely absent as in the leaves of *Panicum colonum*. When it is a membrane it may be broad and oblong, ovate and obtuse, or lanceolate and acute. (See fig. 13.) The function of the ligule is probably to facilitate the shedding of water which may run down the leaf, and thus lessen the danger of rotting of the stem which is sure to follow, if the water were to find its way into the interior of the sheath. Sometimes, in addition to the ligule, other appendages may be present in grass leaves as in *Oryza sativa*. Such outgrowths are called **auricles** or **auricular outgrowths**. (See fig. 13.)

The leaf-blade is well developed in the foliage leaves and in most cases it follows directly on the sheath. But in bamboos and some species of Ischæmum there occurs a short petiole or stalk between the leaf-blade and the sheath. The sheath corresponds morphologically to the leaf base of a leaf of other flowering plants.

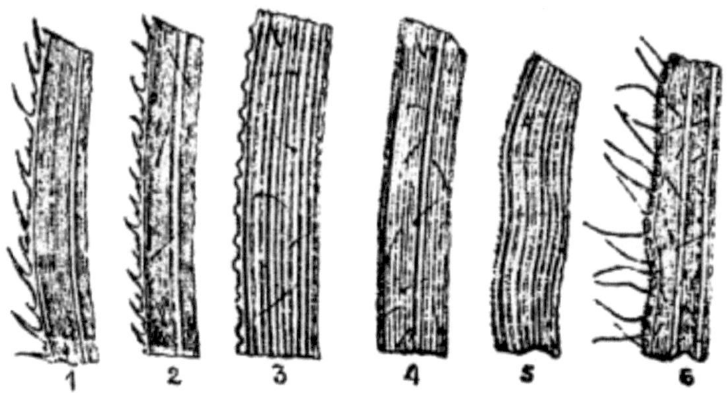

Fig. 15.—Margins of leaves.
1 and 2. Finely serrate; 3. glandular; 4 and 5. very minutely serrate; 6. very minutely serrate and ciliate.

In grasses the leaf-blades usually grow more in length than in any other direction and there is no limit to the length they may attain. Some grasses have very short leaves, others very long ones. The leaf-blade in most grasses is more or less of some elongated form, such as linear, linear-lanceolate, lanceolate, etc. (See fig. 14.) In a few grasses the leaf-blade is ovate, but this is a rare condition. Therefore, in noting the general shape of the leaf-blade the relation of the length to the breadth, the amount of tapering towards the apex and base and the nature of the apex should be considered.

[Pg 12]

The veins in the leaf-blade can usually be seen on holding the leaf up to the light. All the veins run parallel. In most cases the midrib is prominent and in some cases there may be also a distinct keel. Amongst the veins running through the leaf-blade some are large and prominent, while others are small and not conspicuous. On account of this disparity, very often, ridges and furrows become prominent on the upper or lower, or on both the surfaces of the leaf-blades. Generally the two surfaces of the leaf-blade are distinct, and they may be glabrous or hairy. In most grasses the surfaces are rough or scabrid to the touch owing to the presence of regular rows of exceedingly fine sharp pointed minute hairs.

The apex of the blade is generally sharp and pointed, acute or acuminate, or sometimes it may be drawn to a very fine point by gradual tapering. Blunt or obtuse tips are not altogether absent, but it is not a common feature. The leaf-blades in *Panicum colonum* and in some species of Andropogon are rounded or obtuse at the apex.

The margins of the leaf-blade are somewhat hyaline and they may be perfectly even or cut into serrations of fine teeth in various ways. (See fig. 15.) In addition to these minute teeth, there may be long or short cilia. Sometimes the margins are glandular as in *Eragrostis Willdenoviana* and *Eragrostis major*.

The base of the leaf may be narrower, broader than, or about the same as the breadth of the leaf-sheath. It may be rounded, amplexicaul or narrowed. At the base and just above the ligular region there will always be a white distinct zone in the lamina of all grasses called the collar. This collar varies in length and breadth according to the species of grass.

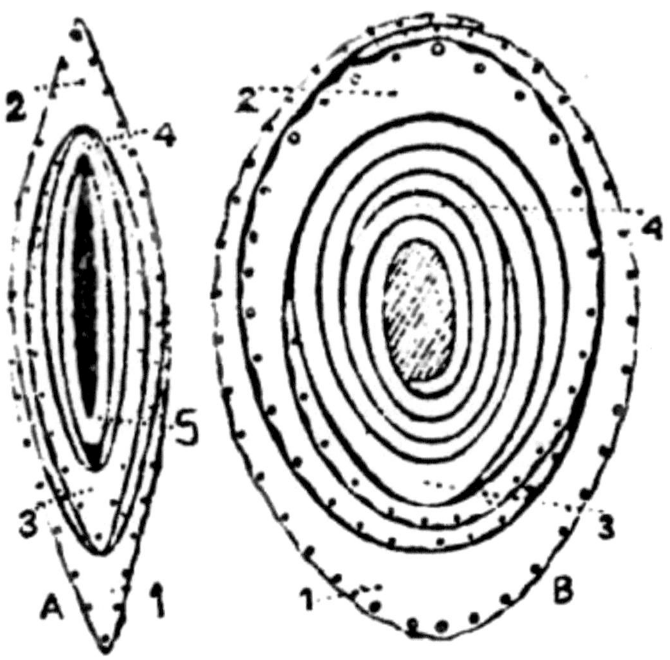

Fig. 16.—Transverse section of leaf-buds.
A. Conduplicate; 1, 2 and 3. leaf-sheaths; 4 and 5. leaf-blades. B.
Convolute; 1 and 2, leaf-sheaths; 3 and 4. leaf-blades.

In young shoots all the leaf-blades are usually found folded at the terminal portions. In most cases the leaf-blade is rolled up inwards from one end to the other so that one margin is inside and the other outside. This folding is termed **convolute**. This is the kind of folding that is found in most grasses. However, there are some grasses such as *Eleusine ægyptiaca* and *Chloris barbata,* in which the folding is different. In these grasses the laminas are folded flat on their midribs so that each half of the blade is folded flat on the other, the inner surfaces being in contact. The leaves are said to be **conduplicate** in this case. When the leaves are conduplicate the shoots are more or less compressed. (See fig. 16.)

[Pg 13]

CHAPTER III.
THE INFLORESCENCE AND FLOWER.

The flowers of grasses are reduced to their essential organs, the stamens and the pistil. The flowers are aggregated together on distinct shoots constituting the inflorescence of grasses. Sooner or later all the branches of a grass-plant terminate in inflorescences which usually stand far above the foliage leaves. As in other flowering plants, in grasses also different forms of inflorescence are met with. But in grasses the unit of the inflorescence is the **spikelet** and not the flower.

The forms of inflorescence usually met with are the spike, raceme and panicle. When the spikelets are sessile or borne directly along an elongated axis as in *Enteropogon melicoides* the inflorescence is a **spike**. If the spikelets borne by the axis are all stalked, however short the pedicels may be, it is a **raceme**. It must, however, be remembered that true spikes are very rare. An inflorescence may appear to be a spike, but on a close examination it will be seen to consist of spikelets more or less pedicelled. Such an inflorescence, strictly speaking, is a **spiciform raceme**. The branches of the inflorescence in *Paspalum scrobiculatum* or *Panicum javanicum* are racemes and the whole inflorescence is a compound raceme. The inflorescence is a **panicle** when the spikelets are borne on secondary, tertiary or further subdivided branches. Panicles differ very much in appearance according to the relative length and stoutness of the branches. In *Eragrostis tremula* the panicle is very diffuse, in *Eragrostis Willldenoviana* less so. The panicle in *Sporobolus coromandelianus* is pyramidal and the branches are all verticillate, the lower being longer than the upper. The branches of a panicle are usually loose, spreading or drooping in most grasses. But in some species of grasses such as *Pennisetum Alopecuros* and *Setaria glauca*, the paniculate inflorescences become so contracted that the pedicels and the short branches are hidden and the inflorescence appears to be a spike. Such inflorescences as these are called **spiciform panicles**. The inflorescences in several species of Andropogon consist of racemes so much modified as to appear exactly like a spike. What looks like a spike in these cases consists of a jointed axis and each joint bears a pair of spikelets, one sessile and the other pedicelled.

The name **rachis** is given to the axis of the spike, raceme and panicle, whether the axis is the main one or of the branch. The rachis of the inflorescence is usually cylindrical. In some grasses it is zigzag as in *Pennisetum cenchroides*. It is very much flattened in *Paspalum scrobiculatum*, but somewhat trigonous in *Digitaria sanguinalis*. In very many grasses the rachis is continuous, but in a few cases it consists of internodes or joints which disarticulate at maturity. Many species of Andropogon have such jointed rachises. Sometimes the joints become greatly thickened [Pg 14] and the surface hollowed out, the spikelets fitting in the cavities as in Rottboellia and Manisuris.

In panicles, especially when they are diffuse, the primary branches may be disposed irregularly or in verticils on the main axis. For example in the panicle of *Eragrostis Willldenoviana*, the branches are irregularly disposed, whereas in *Sporobolus coromandelianus* the branches are verticillate. In both these grasses fleshy cushions are developed in the axils of the branches. These swellings help to spread out the branches especially at the time of anthesis. The branches at the top spread out earlier than those below.

Sometimes at the base of the rachises, main or secondary, glandular streaks are seen as in the rachises of *Sporobolus coromandelianus*. These glands secrete a viscid juice at the time of anthesis.

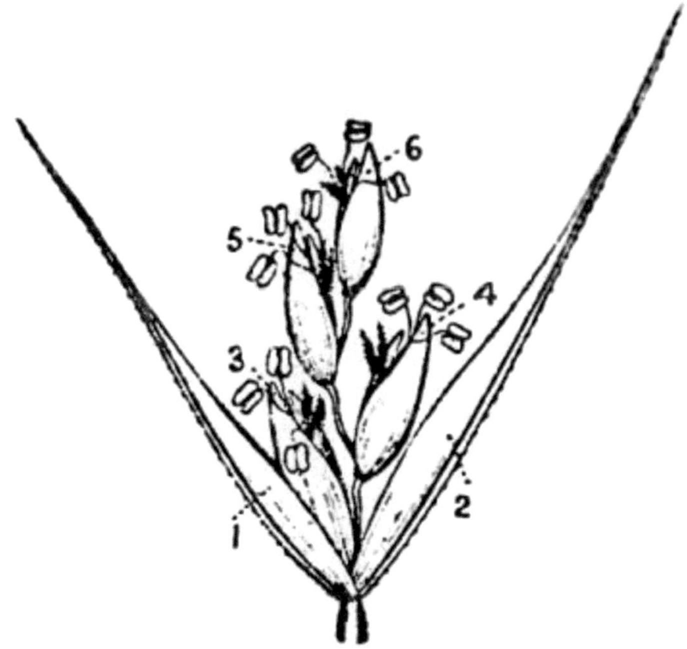

Fig. 17.—The Spikelet of Dinebra arabica.
1 and 2. Empty glumes; 3, 4, 5, and 6. flowering glumes with flowers.

The **spikelet** may be considered as a specialised branch consisting of a short axis, the **rachilla** bearing a series of modified bracts, the **glumes**, the lower pair being empty but the others bearing flowers in their axils. The glumes are two-ranked and imbricating. As a type for the spikelet that of an Eragrostis or Dinebra may be chosen. (See fig. 17.) In this spikelet the rachilla bears a number of glumes alternating and imbricating. The first two glumes at the base of the spikelet do not bear any flowers and so these two glumes are usually called empty glumes. This is the case in almost all the species of grasses. The third and the subsequent glumes are regularly arranged on the slender rachilla alternately in two rows. In the axils of each of these glumes there is a flower, except perhaps in the topmost glume. The flower is usually enclosed by the glume and another structure found opposite the glume and differing very much from the glume. This is the **palea**. It is attached to the axis of the flower and its back is towards the rachilla. Generally there are two nerves in a palea and its margins are enclosed within those of the glume. The palea is homologous with the prophyllum which it very much resembles. The prophyllum is usually found in the branches of grasses, but it is not confined to grasses alone. It occurs in the branches of some species of Commelina.

The spikelets vary very much in their structure. The spikelets in grasses of several genera consist of only four glumes. As usual the first two glumes are empty and the remaining two are flower-bearing glumes. Both these glumes may have perfect flowers as in Isachne or the terminal one may contain a perfect flower, the [Pg 15] lower having either a staminate flower or only a palea. Very often the spikelets are unisexual and the male and female spikelets may be on the same plant as in *Coix Lachryma-Jobi* and *Polytoca barbata*, or they may be on different plants as in *Spinifex squarrosus*.

The glumes of a spikelet are really modified bracts and some differentiate the flowering glumes from the empty ones, by giving them different names. The first two empty glumes are called glumes by all agrostologists. Some in Europe call the flowering glume lower palea to distinguish it from the real palea which they call the upper palea. Some American Authors have recently adopted for the flowering glume the term **lemma** introduced by Piper.

Considerable variation is met with in the case of the empty glumes. Generally they are unequal, the first being smaller. Very often the first glume becomes very small and it may be altogether absent. In some species of Panicum the first glume is very small, in Digitaria it is very minute and in Paspalum and Eriochloa it is entirely suppressed. The flowering glumes are generally uniform when there are many. In the spikelet having only four glumes the fourth glume differs from the others mainly in texture. Instead of being thin and herbaceous it becomes rigid and hard, smooth or rugose externally as in Panicum. Flowering glumes instead of being like empty glumes, become very thin, shorter and hyaline in Andropogon. Sometimes the flowering glumes are awned. All of them may be awned as in Chloris or only the fourth glume as in Andropogon.

The palea is fairly uniform in its structure in many grasses, but it is also subject to variation. It becomes shorter in some and is absent in others. Instead of having two nerves, it may have one and rarely more than two. The palea can easily be distinguished from the glume, because its insertion in the spikelet is different from that of the glume.

Fig. 18.—Flower of Chloris.
1. lodicules; 2. stamens; 3. ovary.

The **lodicules** are small organs and they are the vestiges of the perianth. In most grasses there are only two, but in Ochlandra and other bamboos we meet with three lodicules. There are also some species with many lodicules. In shape they are mostly of some form referable to the cuneate form. They are of somewhat elongated form in Aristida and Chloris. The function of the lodicules seems to be to separate the glume and its palea so as to enable the stamens to come out and hang freely at the time of anthesis. So it is only at the time of the opening of the flowers that the lodicules are at their best. Then they are fairly large, fleshy and thick and conspicuous. In the bud stage they are usually small and after the opening of the flower they shrivel up and are inconspicuous. There are also species of grasses in which the lodicules are not found.

The **stamens** are three in number in the majority of grasses and six are met with in Leersia, Hygrorhiza and Bamboos. Each stamen consists of a very delicate long filament and an anther [Pg 16] basi-fixed to the filament. But as the anthers are long and the connective is reduced to its minimum, they appear as if versatile when the stamens are out. When there are three stamens one stands in front of the flowering glumes and the other two in front of the palea, one opposite each edge of the palea. The relative positions of the parts of the floret are shown in the floral diagrams. (See figs. 18 and 19.)

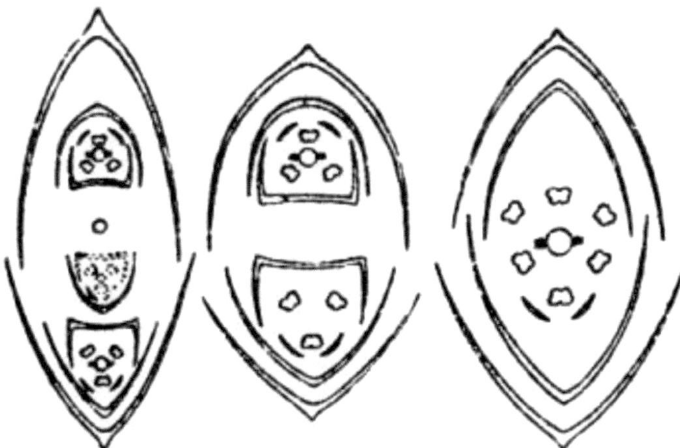

Fig. 19.—Floral diagrams.
The first is that of Chloris, second of Panicum and the third of Oryza.

The **pistil** consists of an ovary and two styles ending in plumose stigmas. The ovary is 1-celled and 1-ovuled. It is one carpelled according to the views of Hackel and his followers and there are also some who consider it as 3-carpelled because of the occurrence of three styles in the pistil of some bamboos.

The **rachilla** is usually well developed and elongated in many-flowered spikelets, while in 1-flowered spikelets it remains very small so that the flower appears to be terminal. It often extends beyond the insertion of the terminal flower and its glume, and then lies hidden appressed to the palea. This may be seen in the spikelets of the species of Cynodon. This prolonged rachilla sometimes bears a minute glume, which is of course rudimentary. Usually the glumes are rather close together on the rachilla so that the internodes are very short; but in some grasses, as in *Dinebra arabica*, the glumes are rather distant and so the internodes are somewhat longer and conspicuous. In some species of Panicum the rachilla is jointed to the pedicel below the empty glumes, whereas it is articulated just above these glumes in *Chloris barbata*. Sometimes the rachilla is articulated between the flowers. This is the case in the spikelet of *Dinebra arabica*.

Pollination in most grasses is brought about by wind, though in a few cases self-pollination occurs. The terminal position of the inflorescence, its protrusion far above the level of the foliage leaves, the swinging and dangling anthers, the abundance of non-sticking pollen and the plumose stigmas are all intended to facilitate pollination by wind. Furthermore the stamens and the stigmas do not mature at the same time. In some grasses the stamens mature earlier, (**protandry**) while in others the stigmas protrude long before the stamens (**protogyny**). As the result of the pollination the ovary developes into a dry 1-seeded indehiscent fruit. The seed fills the cavity fully and the pericarp fuses with the seed-coat [Pg 17] and so they are inseparable. Such a fruit is termed a **caryopsis** or **grain**. Though in the vast majority of grasses the pericarp is inseparable, in a few cases it is free from the seed-coat as in *Sporobolus indicus* and *Eleusine indica*.

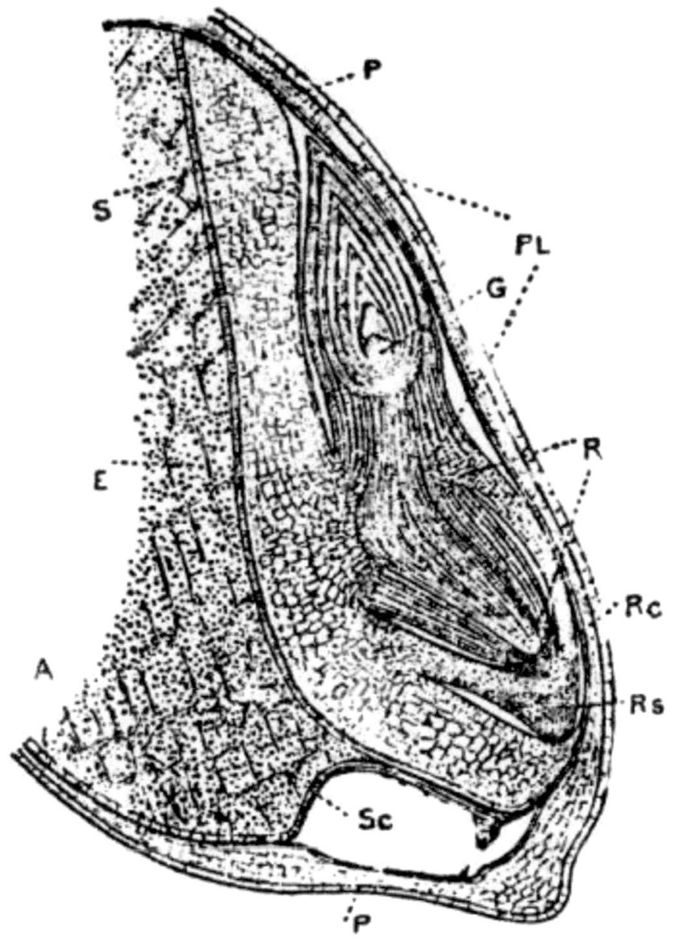

Fig. 20.—Longitudinal section of a portion of the grain of Andropogon Sorghum. × 280

P. Pericarp; Sc. seed-coat; A. aleurone layer; E. endosperm; S. scutellum; Rs. root-sheath; Rc. root-cap; R. radicle; Pl. plumule; G. growing point.

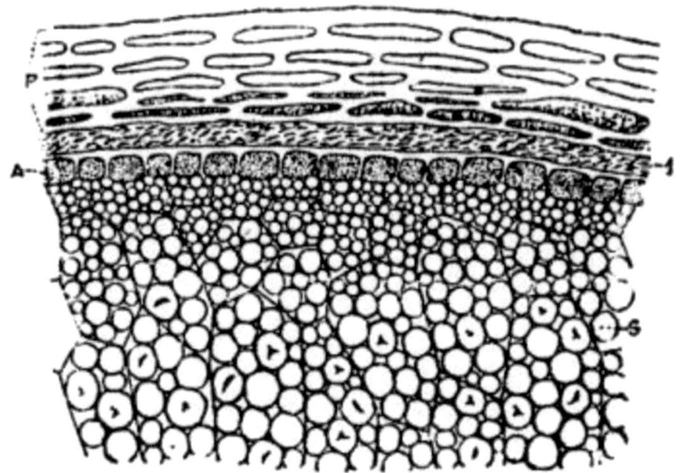

Fig. 21. — A portion of the section of the grain of Andropogon Sorghum. × 500

P. pericarp; I. seed-coat; A. aleurone layer; S. starch.

The caryopsis consists of an embryo on one side at the base and the endosperm occupies the remaining portion. The embryo can be made out on the side of the grain facing the glume, as it is outlined as an oval area. On the other face of the grain which is towards the palea, the hilum is seen at the base. The grain varies in shape considerably. It may be rounded, oval, ellipsoidal, narrow and cylindrical, oblong terete or furrowed. There is considerable variation as regards the colour also.

[Pg 18]

The **embryo** consists of an **axis** and a **scutellum**. The axis, which is differentiated into the plumule directed upward and the radicle downward, is small and straight and it is covered more or less by the edges of the scutellum. The scutellum is attached to the axis at about its middle and its outer surface is in contact with the endosperm. This is an important organ as its function is to absorb nourishment from the endosperm during germination. The scutellum is considered to represent the first leaf or cotyledon. The endosperm consists mostly of starch. Just outside the endosperm and within the epidermis lies a layer of cells containing much proteid substance. This layer is called the **aleurone layer**. (See fig. 21.) As an illustration of the caryopsis, the grain of Andropogon Sorghum may be studied. All the structural details are shown in fig. 20 which is a longitudinal section of the grain.

The primary axis of the embryo is enclosed by a closed sheath both above and below. The sheath which envelopes the radicle is called **coleorhiza** and that of the plumule, **pileole** or **germ-sheath**.

[Pg 19]

CHAPTER IV.
HISTOLOGY OF THE VEGETATIVE ORGANS.

The shoots and roots of grasses conform in their internal structure to the monocotyledonous type. In all grasses numerous threads are found running longitudinally within the stem and some of these pass into the leaves, at the nodes, and run as nerves in the blades of the leaves. These threads are the vascular bundles. The rest of the tissue of the stem and leaves consists of thin-walled parenchymatous cells of different sorts.

The general structure of these bundles is more or less the same in all grasses. A vascular bundle consists of only xylem and phloëm, without the cambium, and so no secondary thickening can take place in the stems of grasses. Such bundles as these are called **closed vascular bundles** to distinguish them from the dicotyledonous type of vascular bundles which are called **open vascular bundles** on account of the existence of the cambium.

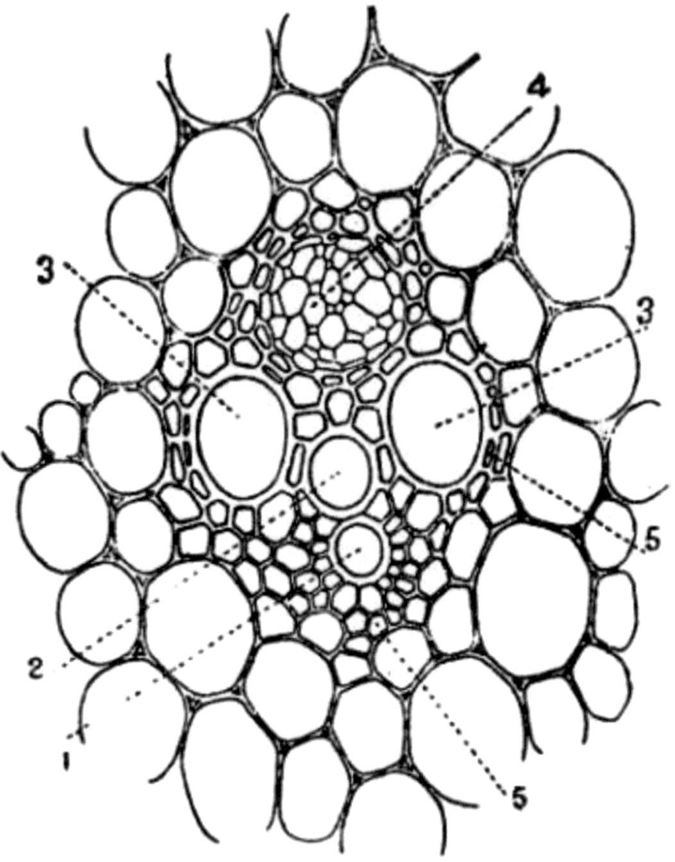

Fig. 22.—Transverse section of a vascular bundle. × 250
1. Annular vessel; 2. spiral vessel; 3. pitted vessel; 4. phloëm or sieve tubes; 5. sclerenchyma.

The component parts and elements of which the vascular bundles in grasses are composed may be learnt by studying the transverse and longitudinal sections of these bundles in any grass. The cross and longitudinal sections of a vascular bundle of the stem of *Pennisetum cenchroides*, are shown in figs. 22 and 23. In the figure of the transverse section the two large cavities indicated by the number 3 and the two small circular cavities with thick walls lying between the larger ones and indicated by the numbers 1 and 2 are the chief elements of the xylem.

By looking at the longitudinal section it is obvious that these elements are really vessels, the larger being pitted and the smaller annular and spiral vessels. These vessels together with the [Pg 20] numerous small thick-walled cells lying between the pitted vessels constitute the xylem. Just above the xylem there is a group of large and small thin-walled cells. This is the phloëm and it consists of sieve tubes and thin-walled cells. All round the xylem and the phloëm there are many thick-walled cells. These are really fibres forming the **bundle-sheath**. On account of this bundle-sheath the bundles are called **fibro-vascular bundles**.

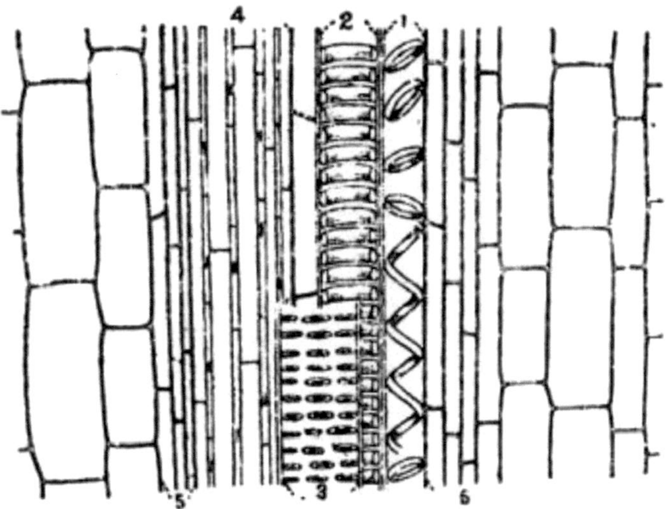

Fig. 23. — Longitudinal section of a vascular bundle. × 250
1. Annular vessel; 2. spiral vessel; 3. pitted vessel; 4. sieve tubes or phloëm; 5. sclerenchyma.

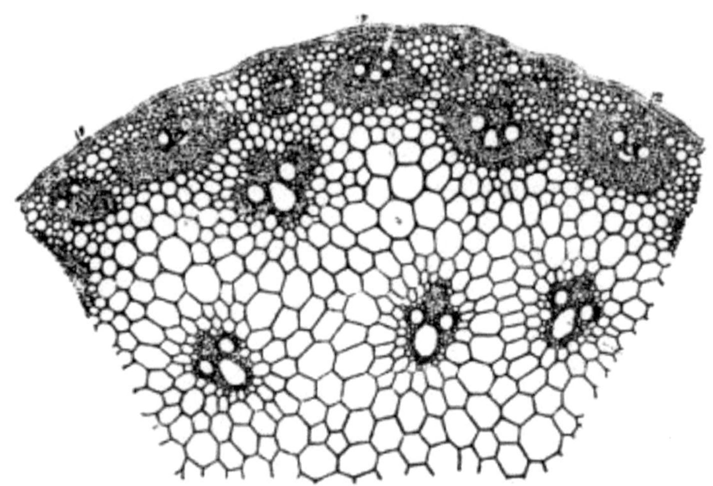

Fig. 24.—Transverse section of a portion of the stem of Rottboellia exaltata. × 70

1. Epidermis; 2. sclerenchyma; 3. vascular bundle.

Structure of the stem.—The stem of a grass consists of a mass of parenchymatous cells with a number of fibro-vascular bundles [Pg 21] imbedded in it, and it is covered externally by a protective layer of cells, the epidermis. The stem is usually solid in all grasses in the young stage, but as it matures the internodes become hollow in many grasses and they remain solid in a few. In the internodes the fibro-vascular bundles run longitudinally and are parallel, but in the nodes they run in all directions and form a net work from which emerge a few bundles to enter the leaves. So far as the broad general features are concerned, the stems of many grasses are more or less similar in structure. However, when we take into consideration the arrangement of bundles, the development and arrangement of sclerenchyma, every species of grass has its own special characteristics. And these are so striking and constant that it may be possible to identify the species from these characters alone.

We may take as a type the stem of *Rottboellia exaltata*. This stem is somewhat semi-circular in transverse section and it is almost straight and flat in the front (the side towards the axillary bud). The peripheral portion of the stem becomes somewhat rigid and thick due to the aggregation of vascular bundles, some small and others large. The outermost series of bundles consisting of small and larger bundles are in contact with the layers of the cells lying just beneath the epidermis and these cells are also thick-walled. A few are away from these being separated by three or four layers of cells from the peripheral bundles. In all these vascular bundles the bundle-sheath is very strongly developed all round and is very much developed especially at the sides. It is this great development of sclerenchyma that makes the outer portion of the cortex hard. Within the ground tissue are found a number of vascular bundles scattered more or less uniformly. These bundles have no continuous bundle-sheaths but have instead groups of fibres at the sides and in front of the phloëm. The cavities near the annular vessels are somewhat larger and conspicuous in these bundles.

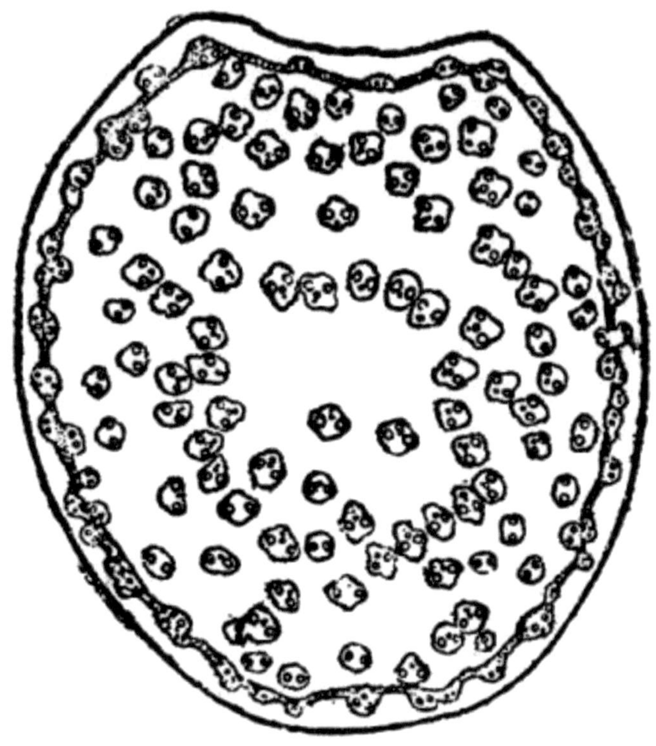

Fig. 25.—Transverse section of the stem of Pennisetum cenchroides. × 20

The epidermal cells are all thickened very much and the outer layer is cutinized and impregnated with silica. This is the case in the epidermis of the stems and leaves of most grasses. (See fig. 24.)

[Pg 22]

In order to give a general idea of the variations in the structure of the stem in grasses a few examples are chosen and the details of the structure of the stems of these grasses are dealt with here.

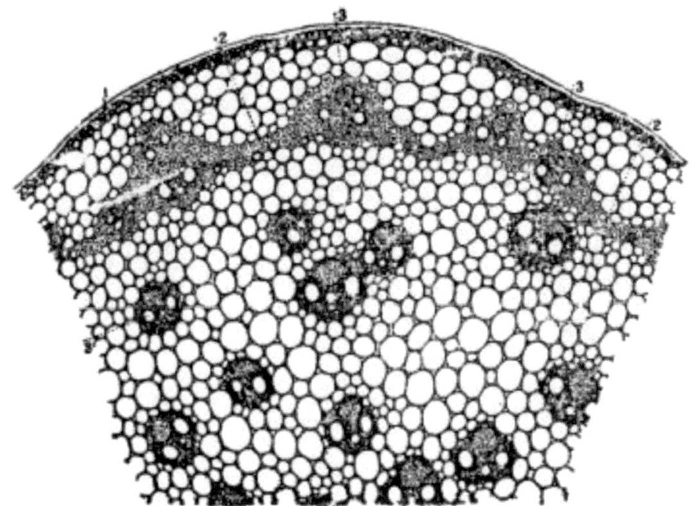

Fig. 26.—Transverse section of a portion of the stem of Pennisetum cenchroides. × 70

1. Epidermis; 2. sclerenchyma; 3. vascular bundle.

The stem of *Pennisetum cenchroides* is somewhat round in outline in the transverse section with a slight curvature in the front. The vascular bundles are rather numerous and irregularly scattered all over the ground tissue. The peripheral bundles are not so close to the periphery of the stem as in *Rottboellia exaltata*. These are separated from the epidermis by several layers of parenchymatous cells. Further, these peripheral bundles are all imbedded in a continuous sclerenchymatous band which runs round the stem in the form of a ring. The epidermal cells as well as the layer of cells in immediate contact with it are thick-walled. In the vascular bundles of the ground tissue the bundle-sheath is rather prominent and the phloëm portion is well developed. (See figs. 25 and 26.)

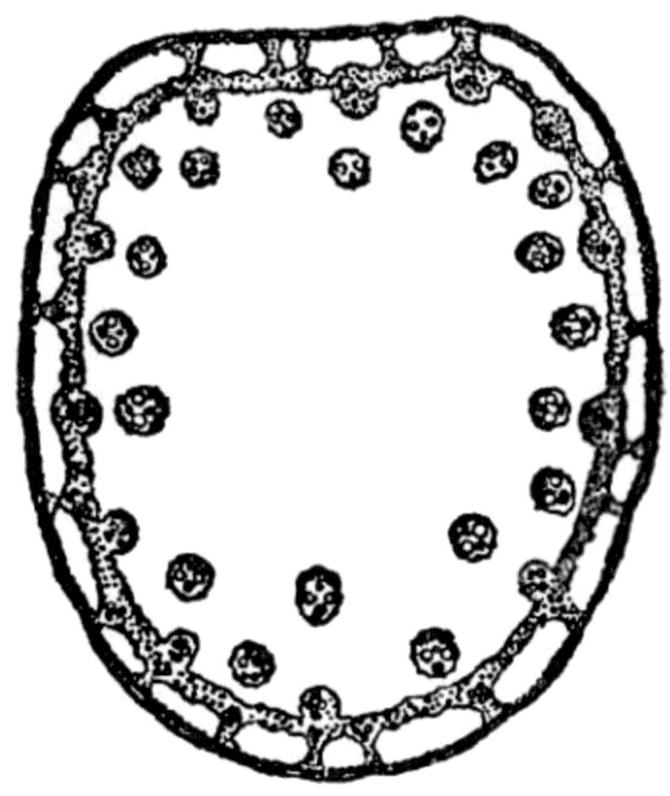

Fig. 27.—Transverse section of the stem of Eriochloa polystachya.
× 25

In the stem of *Eriochloa polystachya*, all the vascular bundles are more or less peripheral in position leaving a wide area of parenchymatous cells in the centre. The outline of the stem in cross section is rotund or ovate-rotund with the front side somewhat flattened and straight. The epidermal cells alone are [Pg 23] thickened. A well developed continuous ring of sclerenchyma is present and this is connected with the epidermal layer at short intervals by means of short sclerenchymatous bands. So the parenchymatous cells of the cortex lying outside the sclerenchymatous ring are di-

58

vided into small isolated areas. There are three series of vascular bundles.

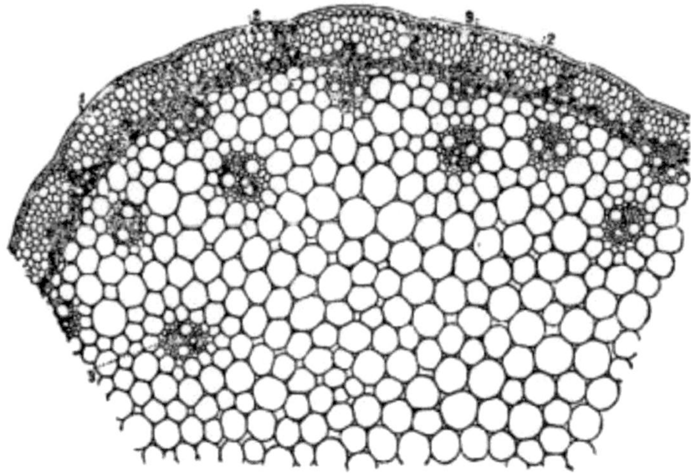

Fig. 28.—Transverse section of a portion of the stem of Eriochloa polystachya. × 70
1. Epidermis; 2. sclerenchyma; 3. vascular bundle.

One series consists of small bundles lying inside the sclerenchyma ring at the base of each of the connecting bands. The second series is made up of large vascular bundles imbedded in the ring so as to bulge out inside the ring. The vascular bundles of the third series are found just away from the ring and separated from it by a few layers of parenchymatous cells. (See figs. 27 and 28.)

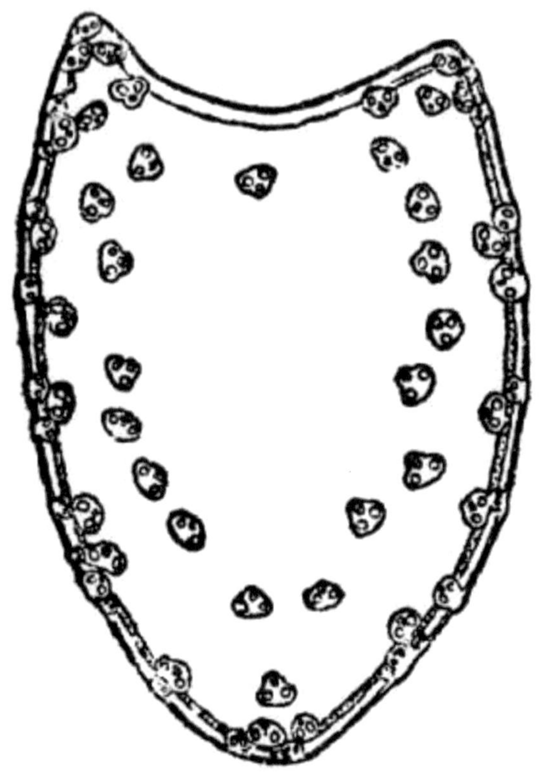

Fig. 29. — Transverse section of the stem of Setaria glauca. × 15

[Pg 24]

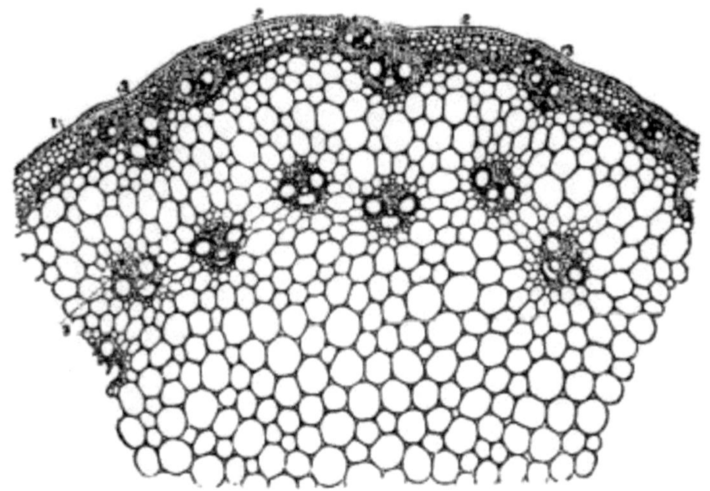

Fig. 30.—Transverse section of a portion of the stem of Setaria glauca. × 50

1. Epidermis; 2. sclerenchyma; 3. vascular bundle.

Another stem in which the vascular bundles are more or less peripheral in position and enclosing a wide parenchyma is that of *Setaria glauca*. In the transverse section of the stem the outline is ovate, laterally compressed, obtusely keeled at the back and somewhat concave in the front. The sclerenchymatous band is narrow and continuous and very close to the epidermis, being separated from it only by two or three layers of thin-walled cells. The epidermal cells alone are thickened. As to the vascular bundles there are three sets. One set of bundles lying just outside the sclerenchymatous ring consists of small ones connecting the ring with the epidermis. Just inside the sclerenchymatous ring lies a series of bundles which are connected with it. Still inside, at some distance from the sclerenchymatous band, are seen vascular bundles forming a row and enclosing a large space of the ground tissue consisting of only parenchyma. (See figs. 29 and 30.)

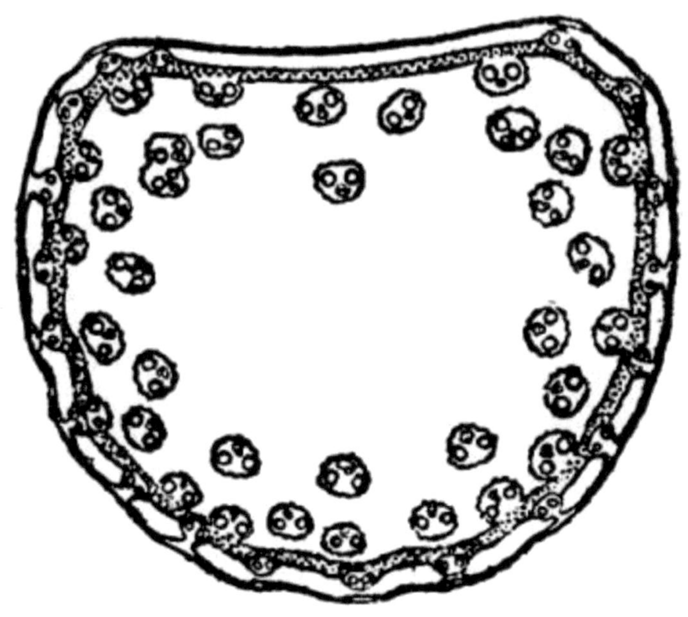

Fig. 31.—Transverse section of the stem of Panicum ramosum ×
24

[Pg 25]

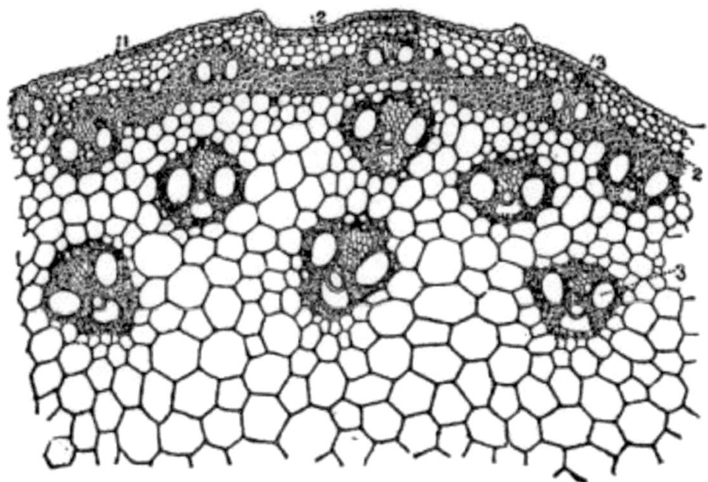

Fig. 32. — Transverse section of a portion of the stem of Panicum ramosum. × 75

1. Epidermis; 2. sclerenchyma; 3. vascular bundle.

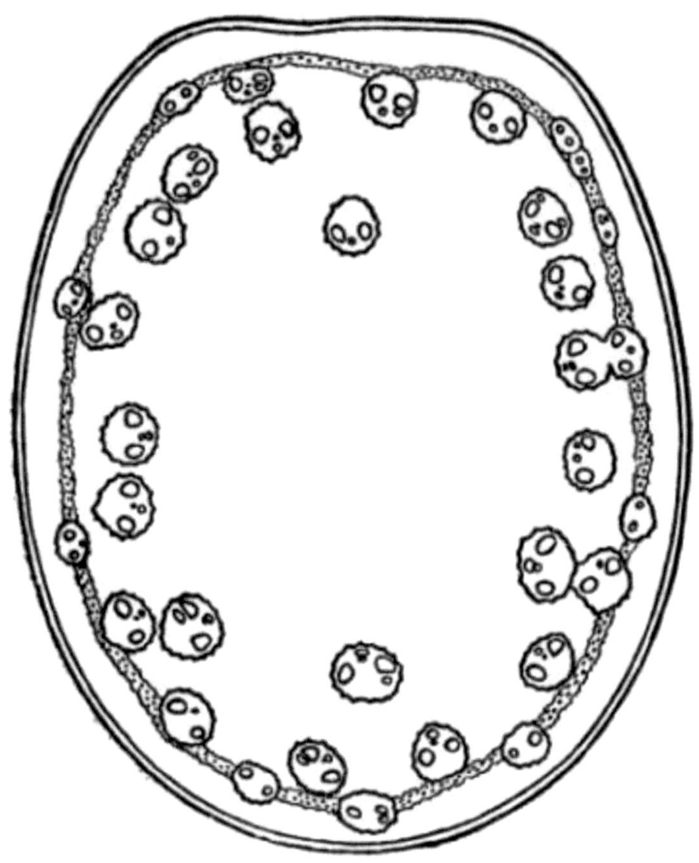

Fig. 33.—Transverse section of the stem of Andropogon caricosus.
× 25

The stem of *Panicum ramosum* is semi-circular and somewhat flat
on one side. The epidermal cells alone are thickened. There [Pg 26]
is a broad well developed continuous band of sclerenchyma, which
is connected at regular intervals with the epidermis by small vascular bundles. Another row of vascular bundles lies just inside the
sclerenchymatous ring and each of these bundles is in contact with
the band. Away from the ring lie a number of bundles forming a

series disposed in two irregular rings around a broad portion of the ground tissue. (See figs. 31 and 32.)

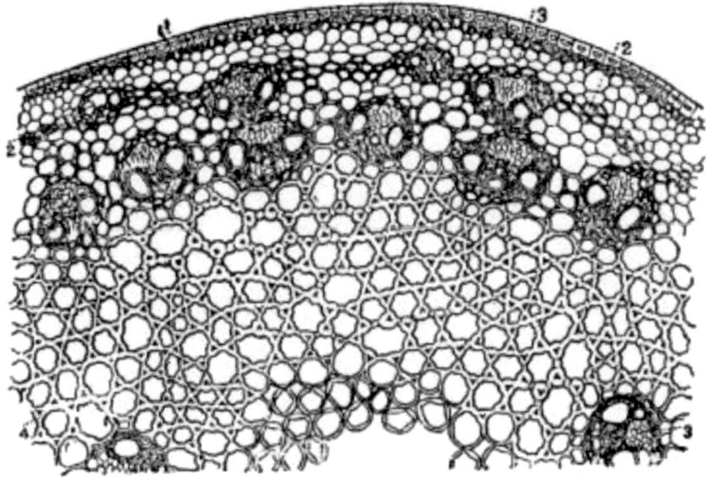

Fig. 34.—Transverse section of a portion of the stem of Andropogon caricosus. × 75
1. Epidermis; 2. sclerenchyma; 3. vascular bundle.

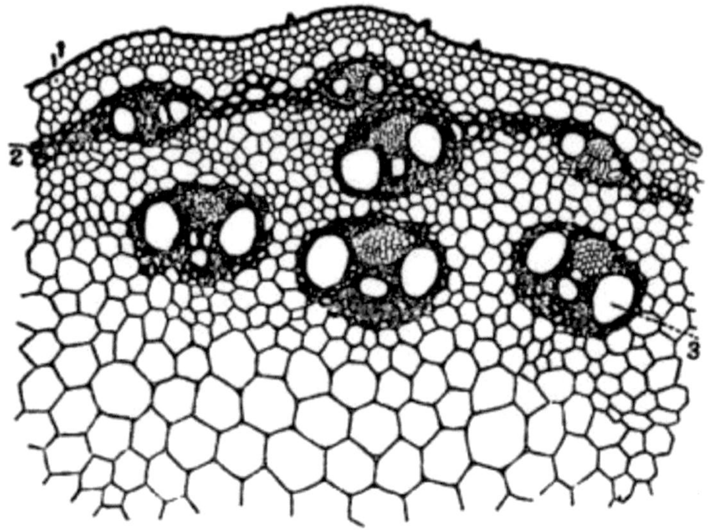

Fig. 35.—Transverse section of a portion of the stem of Panicum Isachne. × 100

1. Epidermis; 2. sclerenchyma; 3. vascular bundle.

The stem of the grass *Andropogon caricosus* is oval in outline, the front being flat. The epidermal cells and those below and in contact with them are thick-walled. The sclerenchymatous ring though present is very narrow and not very conspicuous. It consists of one or two layers of cells connecting a few vascular bundles forming the outermost set. There is a series of vascular bundles inside the ring which surrounds a large area of the ground tissue. Two isolated bundles, one in front and another at the back of the ground tissue, are found. The [Pg 27] cells of the ground tissue lying just inside the vascular bundles are all very much thickened. (See figs. 33 and 34.)

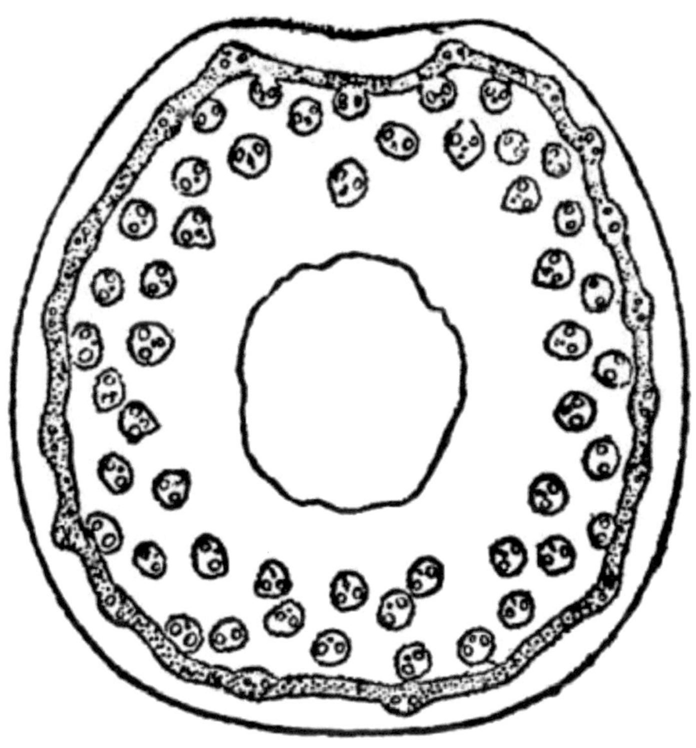

Fig. 36. — Transverse section of the stem of Eragrostis interrupta. ×
25

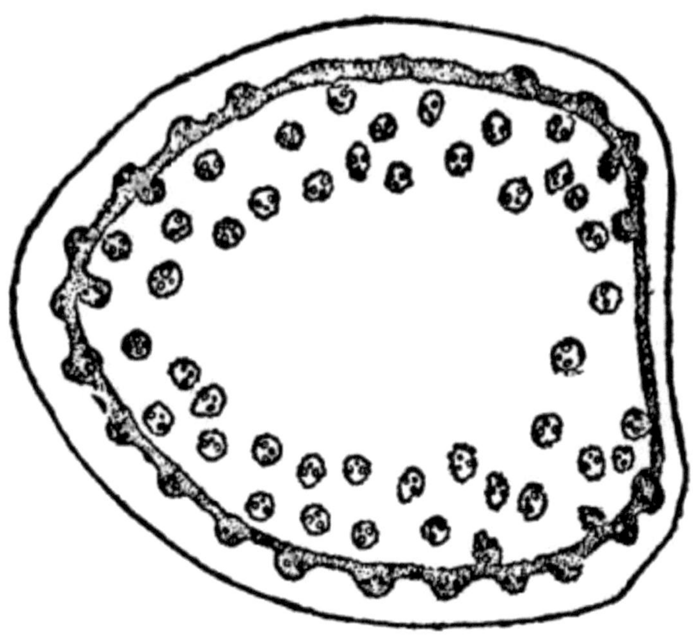

Fig. 37.—Transverse section of the stem of Panicum flavidum. ×
15

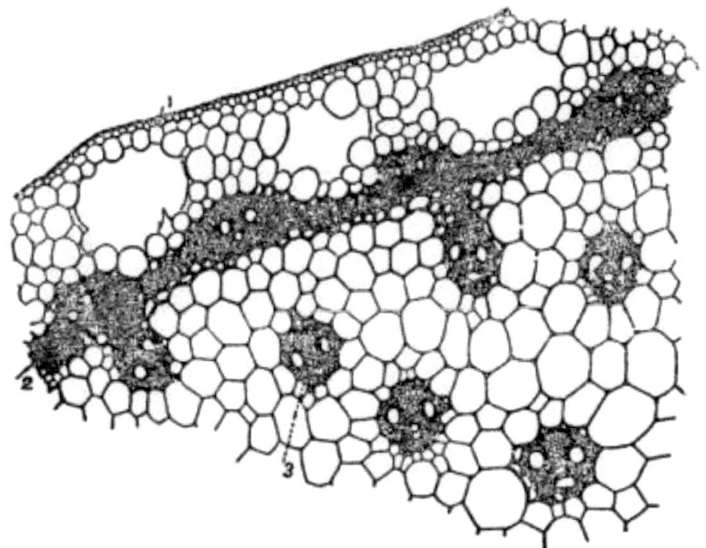

Fig. 38.—Transverse section of a portion of the stem of Panicum flavidum. × 70
1. Epidermis; 2. sclerenchyma; 3. vascular bundle.

The stems of *Panicum Isachne* and *Eragrostis interrupta* are hollow. The stem of the former is circular in outline in cross section, though wavy. There is a sclerenchymatous ring close to the epidermis but separated from it by a few layers of parenchyma. One set of bundles is imbedded in the band, and another set just touches the inner border of it. A third series is disposed around a fairly large amount of ground tissue, which may or may not have a cavity in the centre. The stem of *Eragrostis* [Pg 28] *interrupta* has more or less the same structure, but the cortex has air spaces here and there. Other minor differences may be seen on referring to figs. 35 and 36.

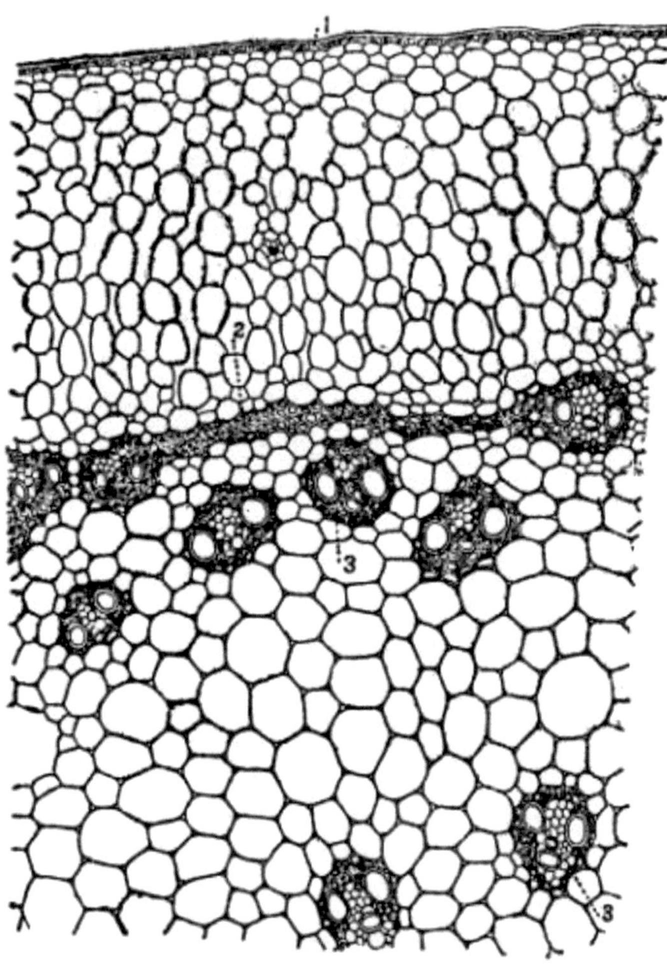

Fig. 39.—Transverse section of a portion of the stem of Panicum colonum. × 70

1. Epidermis; 2. sclerenchyma; 3. vascular bundle.

The stems of grasses growing in wet or marshy situations differ in structure from those detailed above. As examples the stems of *Panicum flavidum, Panicum colonum, Panicum Crus-galli* and *Panicum fluitans* may be considered. The stem of *Panicum flavidum* is broadly ovate in cross section with a flat front and is more or less solid, though occasionally the parenchymatous cells in the centre get broken. Two rows of vascular bundles surround a fairly large amount of parenchymatous cells of the ground tissue. There is a continuous ring of sclerenchyma separated from the epidermis by a fairly broad cortex. The cortex has a number of fairly large air-cavities separated by bands of parenchymatous cells. Within the sclerenchymatous band lie small vascular bundles at regular intervals just towards the cortex. A few isolated bundles are in contact with the inner border. (See figs. 37 and 38.)

The stems of *Panicum colonum, Panicum stagninum* and *Panicum Crus-galli* have in their centre in the ground tissue stellate cells with air-cavities. This part is surrounded by a fairly broad portion of parenchymatous cells in which are imbedded two rows [Pg 29] of bundles. Outside these bundles runs round the stem a narrow sclerenchymatous band with a few bundles in it of which some touch it inside and others outside. Two bundles are found by themselves in the tissue of stellate cells. In *Panicum Crus-galli* three or four bundles are met with amidst the stellate cells.

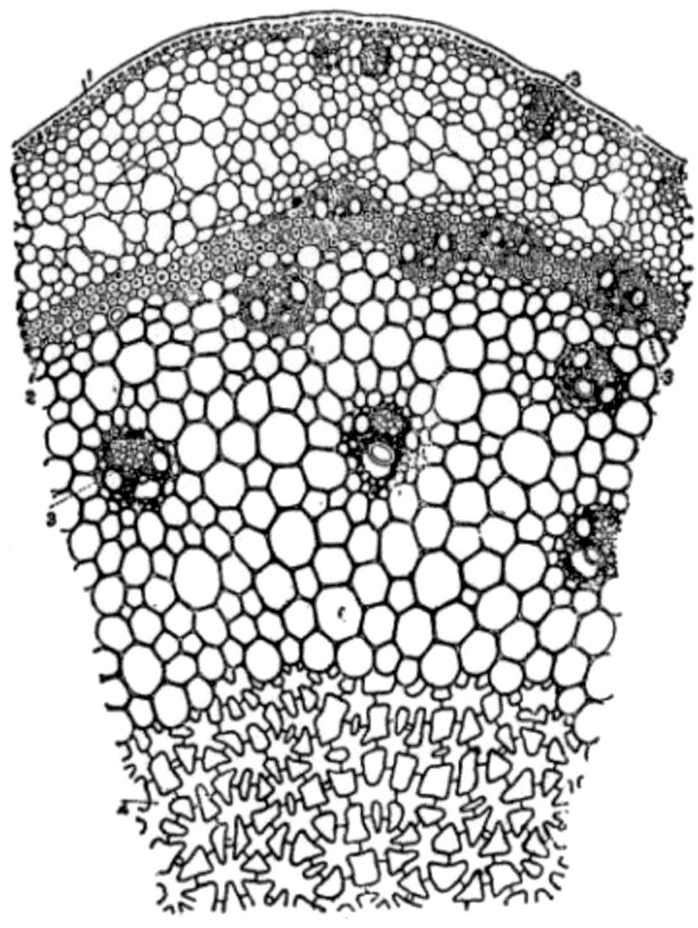

Fig. 40. — Transverse section of a portion of the stem of Panicum
Crus-galli. × 70
1. Epidermis; 2. sclerenchyma; 3. vascular bundle.

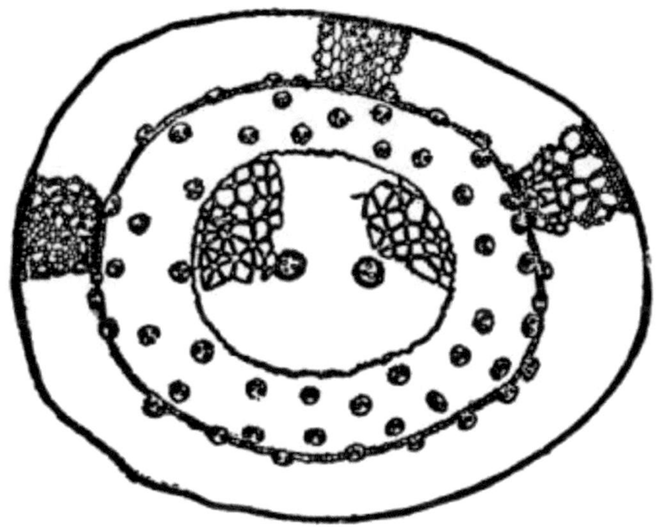

Fig. 41.—Transverse section of the stem of Panicum stagninum. ×
10

The cortex outside the band of sclerenchyma is full of air-cavities, small and large. In *Panicum colonum* the outline of the [Pg 30] stem is ellipsoidal with the front quite flat, and the cortex is narrow at the sides and very broad in front and at the back. The sclerenchymatous ring is circular in outline. The stem of *Panicum Crus-galli* is broadly ovoid and the cortex is uniformly broad. The epidermal cells as well as the lower cells are thickened in the stems of *Panicum fluitans* and *Panicum Crus-galli*, but in the stems of *Panicum colonum* and *Panicum flavidum* the epidermis alone is thickened. In the cortical portion outside the sclerenchymatous band, small vascular bundles occur in the stems of *Panicum colonum*, *Panicum Crus-galli* and *Panicum fluitans*. (See figs. 39, 40, 42 and 43.)

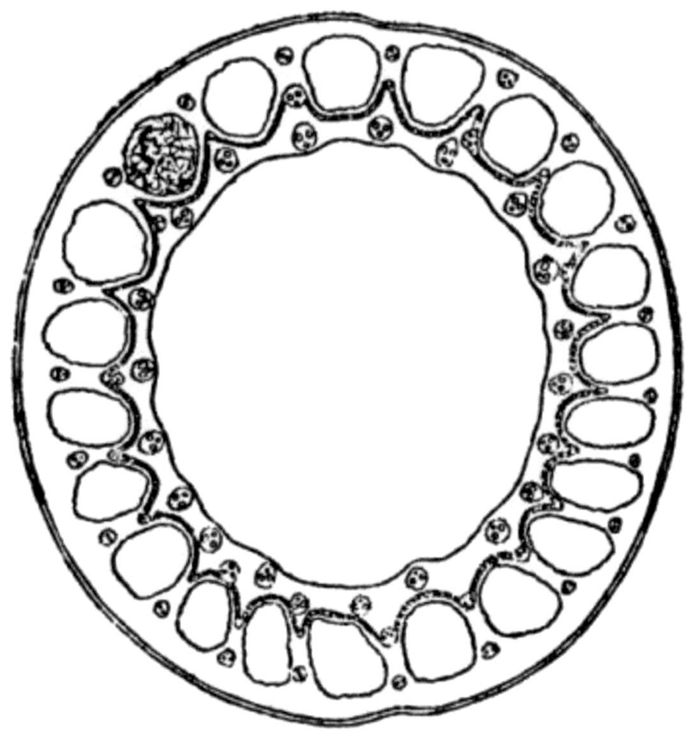

Fig. 42.—Transverse section of the stem of Panicum fluitans. × 15

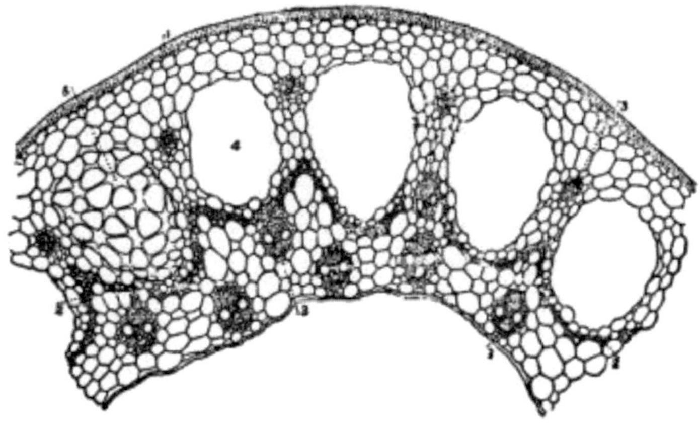

Fig. 43.—Transverse section of a portion of the stem of Panicum fluitans. × 60
1. Epidermis; 2. sclerenchyma; 3. vascular bundle; 4. air-cavity; 5. diaphragm.

[Pg 31]

The stem of *Panicum fluitans* is round in outline in the transverse section and has a large cavity. Just close to the cavity and separated from it by only one or two parenchymatous cells are found vascular bundles forming a series. Outside this series of bundles lies a sclerenchymatous band which is wavy, following the lower edges of the large air-cavities. One series of bundles is connected with this sclerenchymatous ring. The air-cavities are large and uniform and are separated by bands of parenchymatous cells. In each of these bands lies a vascular bundle on the upper side near the periphery. Sometimes we find, especially in young stages, diaphragms of stellulate cells stretched across the air-cavities. Later as the stem matures these disappear and the cavities become conspicuous. (See figs. 42 and 43.)

Structure of the root.—As already stated, the roots of grasses conform to the monocotyledonous type, but the variations met with in their structure are not so great as in the case of the stem. The root-tips are protected by root-caps, and the actual tip of the root is very distinct in the roots of all grasses and it can be seen very clearly in a longitudinal section of the root. The actual tip of the root is sharply distinct from the root-cap as there are two distinct sets of cells, one giving rise to the root-tip and the other to the root-cap.

The young root-tips are always free from root-hairs, and they are confined to the portions behind the root-tips. The extent of the root-hair region will vary according to the vigour and development of the roots and the nature of the soil. The root-hairs are mere protrusions of the cells of the outermost layer of the cortex of the root and this layer is called the **piliferous layer**.

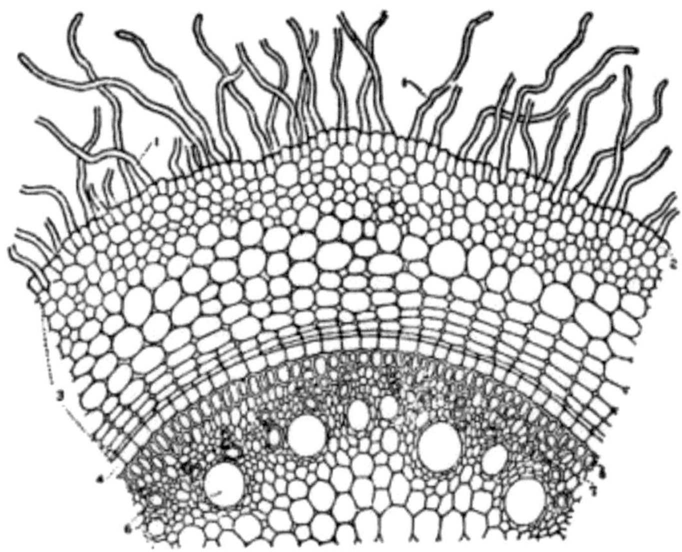

Fig. 44.—Transverse section of a part of the root of Pennisetum cenchroides. × 100
1. Root-hair; 2. piliferous layer; 3. cortex; 4. endodermis; 5. pericycle; 6. xylem; 7. phloëm.

[Pg 32]

To learn the structure of the roots of grasses we may select as types the roots of *Pennisetum cenchroides* and *Andropogon Sorghum* and consider their structural details. In the transverse sections of these roots we find a fairly broad cortex consisting of thin-walled parenchymatous cells more or less regularly arranged. (See figs. 44

and 45.) Just below the piliferous layer two or three layers of thick-walled cells are seen. In the roots of *Andropogon Sorghum* these thick-walled cells are very conspicuous as they consist of several layers. These layers of thick-walled cells constitute the **exodermis**. (See fig. 46.) The innermost layer of cells of the cortex is called the **endodermis** and it becomes conspicuous on account of the thickening in the lateral and inner walls of the cells of this layer. (See figs. 44 and 47.)

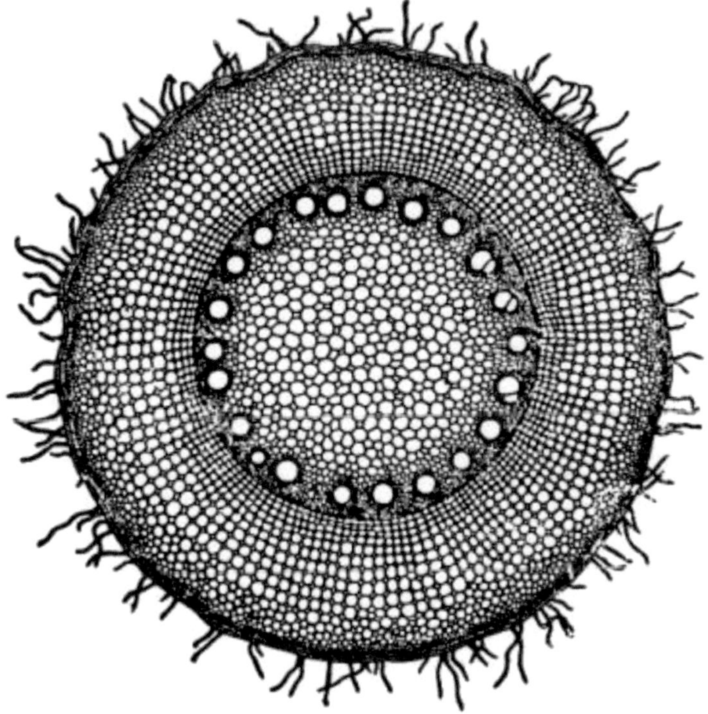

Fig. 45.—Transverse section of the entire root of Andropogon Sorghum. × 25

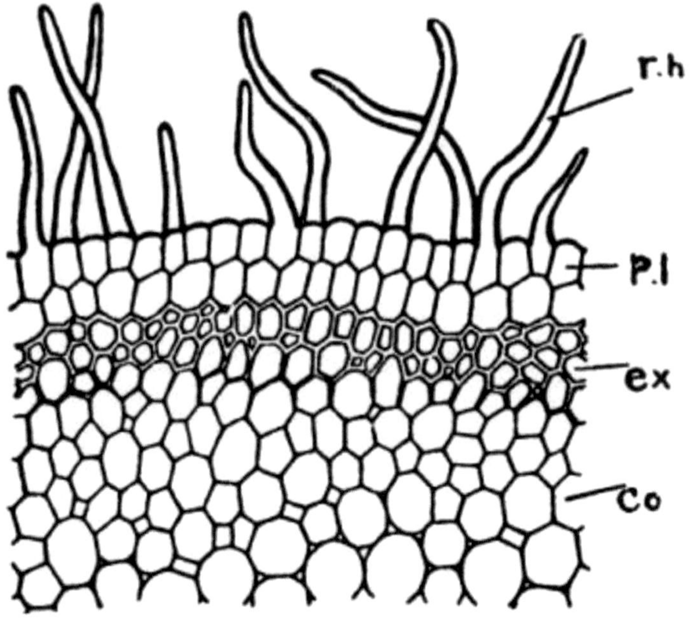

Fig. 46.—Transverse section of the cortical portion of the root of Andropogon Sorghum. × 150
r.h. Root-hair; p.l. piliferous layer; ex. exodermis; co. cortex.

The rest of the root forming the central core is the stele and at its periphery there is a single layer of cells called the **pericycle**. The arrangement of the xylem and the phloëm is different from that of the stem. They lie side by side on different radii, and not one behind the other on the same radius as in the stem. The number of xylem groups is fairly large and the development of the xylem is from the pericycle towards the centre of the stele. (See figs. 44 and 45.) The parenchymatous cells in the centre of the stele become thick-walled in older roots.

Structure of the leaf.—The structure of the leaf of grasses is quite characteristic of the family. In every leaf a number of [Pg 33] vascular bundles, some small and others large, pass from the base to the apex. Externally the leaf is covered on both the sides by the epidermis. The spaces existing between the vascular bundles and the epidermis are filled with parenchymatous cells. The larger vascular bundles consist of xylem and phloëm surrounded by a bundle sheath of a single layer of cells. In the smaller bundles the xylem is very much reduced. Around every vascular bundle there is a single row of somewhat large cells densely packed with large chloroplasts, the **chlorophyllous layer**. The vascular bundles are strengthened by fibres, on both the sides in the case of larger bundles and on only one side in small bundles.

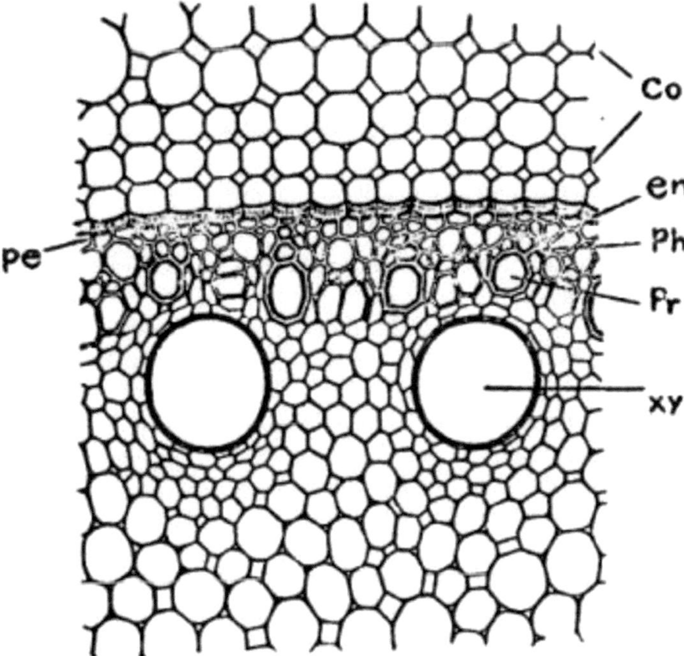

Fig. 47.—Transverse section of the stele portion of the root of Andropogon Sorghum. × 150
Co. Cortex; en. endodermis; pe. pericycle; ph. phloëm; pr. protoxylem; xy. xylem vessel.

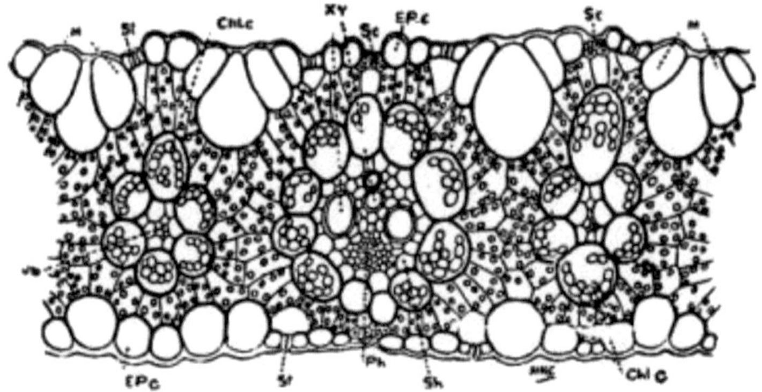

Fig. 48.—A portion of the transverse section of the leaf of Panicum javanicum. × 100

Ep. c. An ordinary epidermal cell; st. stomata; sc. sclerenchyma; ph. phloëm; chl c. chlorenchyma; m. motor cells; xy. xylem.

For a detailed study of the structure of the leaves of grasses the leaf of the grass *Panicum javanicum* may be chosen. In a transverse section of this leaf, the vascular bundles are very conspicuous. The larger bundles are normal in every way, while in the smaller ones the xylem elements are considerably reduced. Around every one of the vascular bundles there is a single row of large cells containing large chlorophyll grains (the chlorophyllous layer). In a well developed large vascular bundle the chlorophyllous layer is open below just close to the sclerenchymatous band. On both sides of the larger vascular bundle there are bands of sclerenchyma. In the case of smaller bundles some are strengthened by sclerenchyma on the lower side and others have none. The spaces between the bundles are occupied by thin-walled parenchymatous cells containing small chlorophyll grains.

[Pg 34]

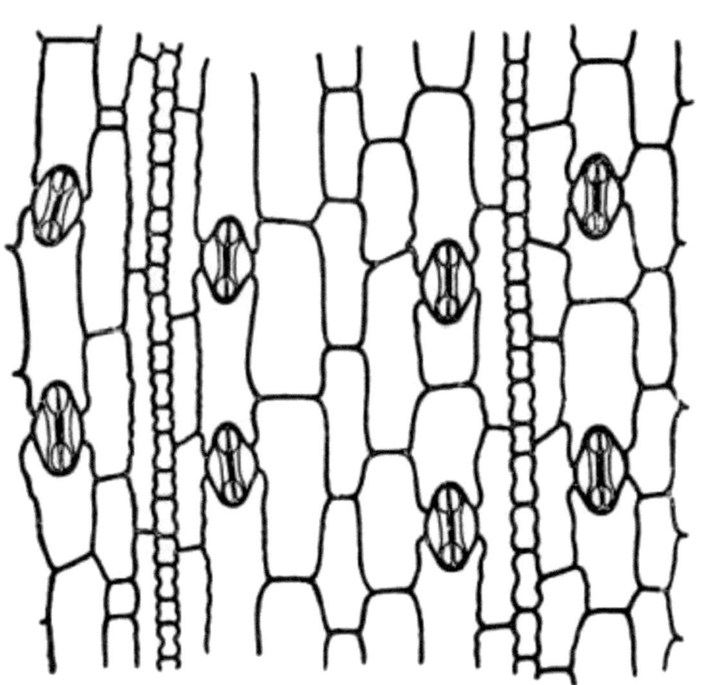

Fig. 49.—Upper epidermis of the leaf of Panicum javanicum. ×
300

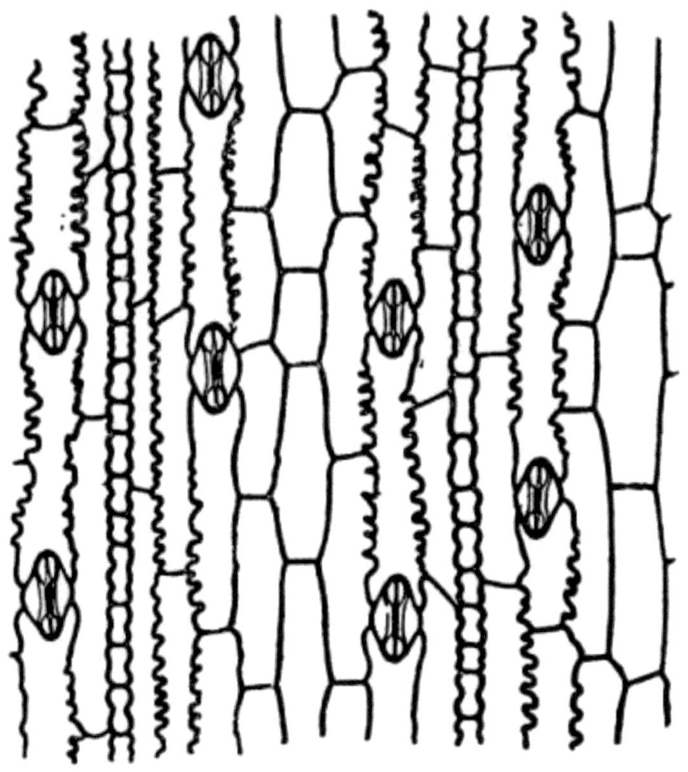

Fig. 50.—Lower epidermis of the leaf of Panicum javanicum. ×
300

The lower epidermis of the leaf in the transverse section is even
and consists of small and large round cells. The upper epidermis is
slightly wavy and it is made up of some small round cells alternating with groups of larger cells. The epidermal cells lying over sclerenchyma and the smaller vascular bundles are small [Pg 35] and
round, while those lying over the furrows between the vascular
bundles are large and are called **motor** or **bulliform cells**. The presence of motor cells is a characteristic feature of the leaves of many
grasses.

The continuity of both the upper and the lower epidermis is interrupted by the stomata. Air-cavities are seen below these stomata. The arrangement of the stomata, the shape of the guard cells and the characteristics of the epidermal cells become clear on examining a piece of epidermis. (See figs. 49 and 50.)

The structure of the leaf of *Panicum javanicum* may be taken as typical of the structure of the leaves of most grasses. The leaves of *Eriochloa polystachya*, Cynodon and Paspalums are very much like the leaves of *Panicum javanicum* in their internal structure.

Considerable amount of variation, however, occurs in the leaves of grasses especially as regards the arrangement of fibres and motor cells.

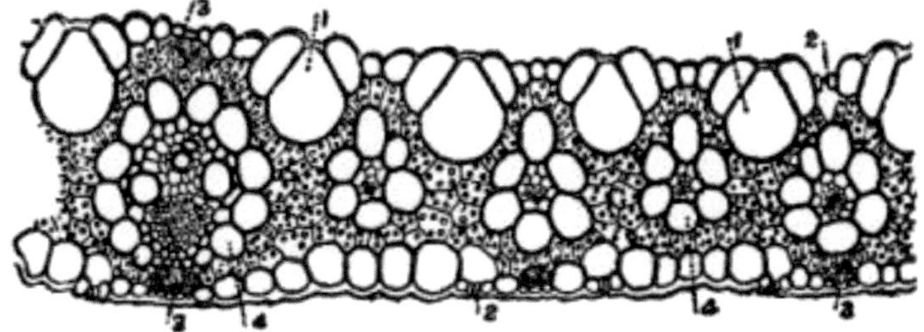

Fig. 51.—A portion of the transverse section of the leaf of Eriochloa polystachya × 120
1. Motor cell; 2. stomata; 3. sclerenchyma; 4. chlorophyllous layer.

Every large primary vascular bundle in the leaves of many grasses possesses sclerenchymatous bands both above and below. The other vascular bundles may have bands of sclerenchyma on both sides or on one side only or none. For example, in the leaves of *Panicum repens* both the primary and secondary bundles are provided with sclerenchyma on both the sides, while those of the third order may have it on one side or not. The hyaline margin of this leaf and of the leaves of other grasses consists entirely of sclerenchyma. (See fig. 53.)

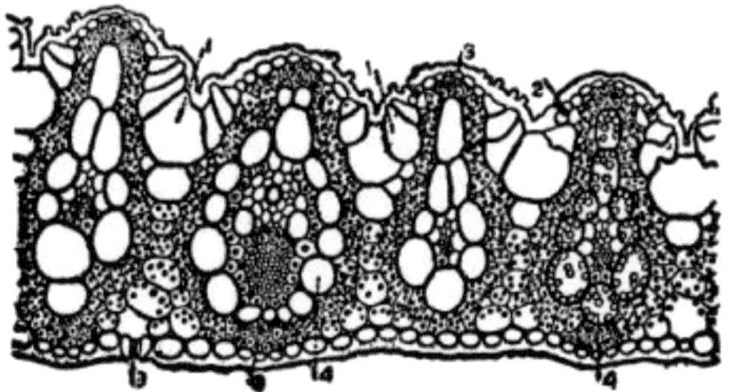

Fig. 52.—Transverse section of a portion of the leaf of Panicum repens. × 120
1. Motor cells; 2. stomata; 3. sclerenchyma; 4. chlorophyllous layer.

[Pg 36]

All the vascular bundles in the leaves of *Aristida setacea* have broad sclerenchymatous bands on both the sides. Besides these bands arranged like a girder above and below each bundle, there are on the lower side bands of sclerenchyma. So the sclerenchyma becomes almost continuous on the lower side.

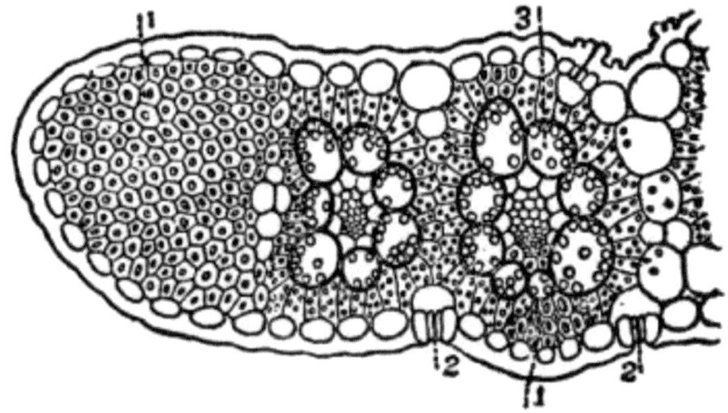

Fig. 53.—Transverse section of the leaf margin of Panicum repens. × 180
1. Sclerenchyma; 2. stomata; 3. chlorophyllous layer.

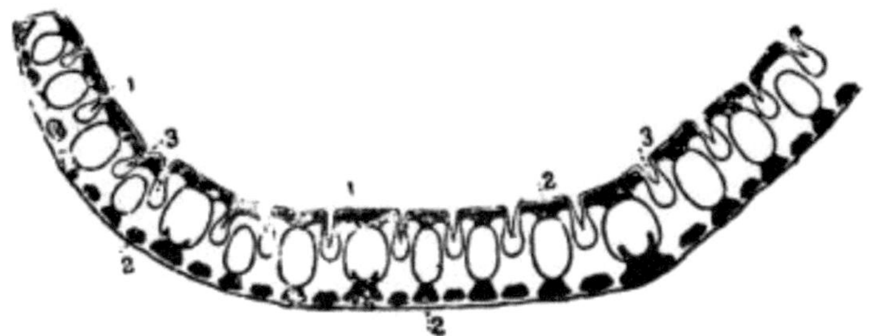

Fig. 54.—Transverse section of a part of the leaf of Aristida setacea.
× 30.
1. Vascular bundle; 2. sclerenchyma; 3. motor cells.

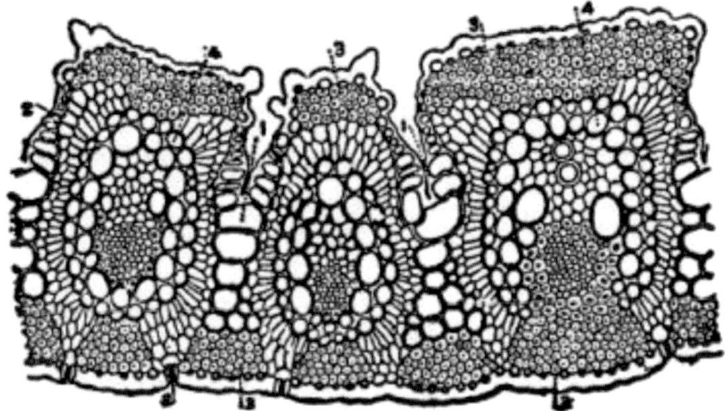

Fig. 55.—Transverse section of a portion of the leaf of Aristida setacea. × 120
1. Motor cells; 2. stomata; 3. sclerenchyma; 4. epidermis; 5. cutin
layer.

The sclerenchyma lying on the lower side of the primary bundles
are contiguous with the bundle, while those above are separated
from the bundle by the chlorophyllous layer. (See fig. 55.) In the
case of secondary and tertiary bundles the sclerenchymatous [Pg 37]
bands lying on the lower side are in contact with the chlorophyllous
layer, whereas the upper bands are either in contact with this layer
or separated from it by a few parenchymatous cells.

All the vascular bundles in the leaves of *Eragrostis Willldenoviana* are provided with sclerenchyma on both the sides. The lower band of the primary vascular bundles is continuous with the vascular bundle, the chlorophyllous layer being open below. The upper bands of the primary and the lower bands of the secondary vascular bundles just touch the chlorophyllous layer. In the secondary bundles the sclerenchyma band above is separated from the chlorophyllous layer by two layers of parenchyma. In the case of the leaves of *Panicum flavidum, P. colonum, P. fluitans* and *Pennisetum cenchroides* the sclerenchyma is separated from the chlorophyllous layer by layers of parenchyma.

Fig. 56.—Transverse section of a part of the leaf of Eragrostis Willdenoviana. × 30
1. Vascular bundle; 2. sclerenchyma; 3. motor cells.

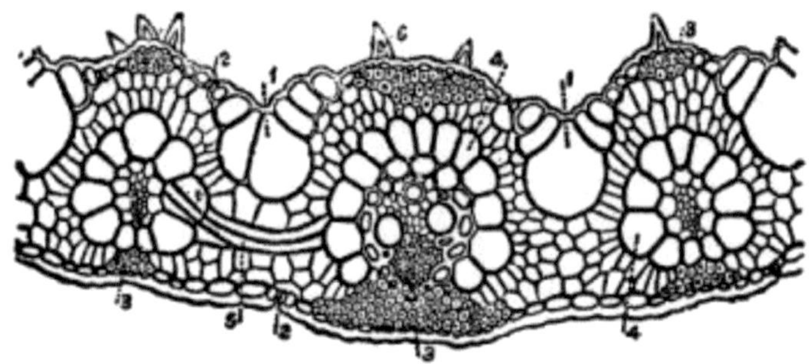

Fig. 57.—Transverse section of a portion of the leaf of Eragrostis Willdenoviana. × 150
1. Motor cells; 2. stomata; 3. sclerenchyma; 4. chlorophyllous layer; 5. vascular strand cut through; 6. hair.

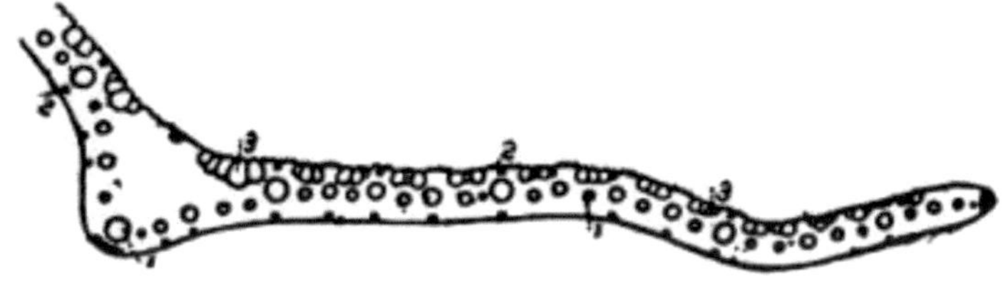

Fig. 58.—Transverse section of a part of the leaf of Panicum colonum. × 30

1. Vascular bundle; 2. sclerenchyma; 3. motor cells.

Even from the few examples dealt with above, it is obvious that the range of variation of sclerenchyma in leaves is very great. In the leaves of *Aristida setacea* there is a considerable amount of sclerenchyma whilst in some leaves such as those of *Panicum* [Pg 38] *colonum, P. flavidum* and *Panicum fluitans* the sclerenchyma is reduced to its minimum.

Fig. 59.—Transverse section of a part of the leaf of Panicum fluitans. × 30

1. Vascular bundle; 2. sclerenchyma.

In the leaves of grasses growing in dry situations the development of sclerenchyma is generally very considerable. The grass *Aristida setacea* is a good example of a xerophytic grass. The seashore grass *Spinifex squarrosus* is another example of the same kind. But in the leaves of this grass, the development of sclerenchyma is not very considerable, but there is a great development of parenchymatous cells free from chlorophyll within the leaf, the chlorophyll bearing cells being confined to the upper and the lower surfaces of the leaves.

Fig. 60.—Transverse section of a leaf of Spinifex squarrosus. × 10
1. Vascular bundle; 2. sclerenchyma.

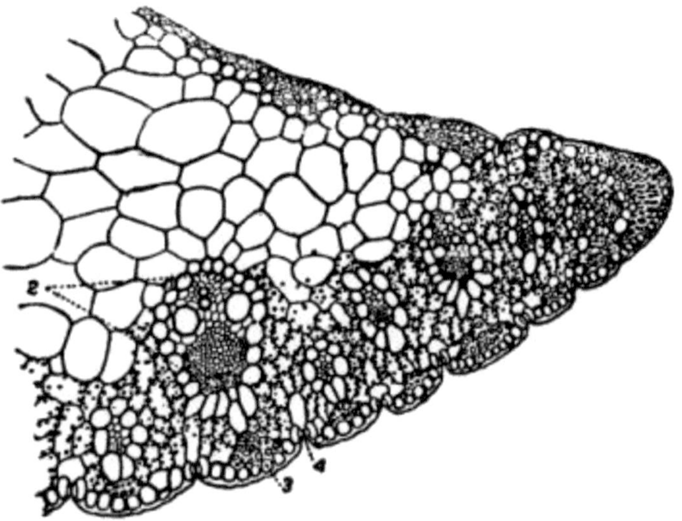

Fig. 61.—A portion of the transverse section of the leaf of Spinifex squarrosus. × 60
1. Sclerenchyma; 2. vascular bundle; 3. epidermis; 4. stomata.

[Pg 39]

The upper and the lower surfaces of the leaves of many grasses are more or less even, but in the case of a few grasses the upper surface consists of ridges and furrows, instead of being even. In the leaves of *Panicum repens* and *Eragrostis Willdenoviana* the upper surface is wavy and consists of shallow furrows and slightly raised ridges. But in the leaves of *Aristida setacea* and *Panicum fluitans* the furrows are deeper and the ridges are more prominent. In *Aristida setacea* the ridges are flat-topped and they are rounded with broad furrows in *Panicum fluitans*.

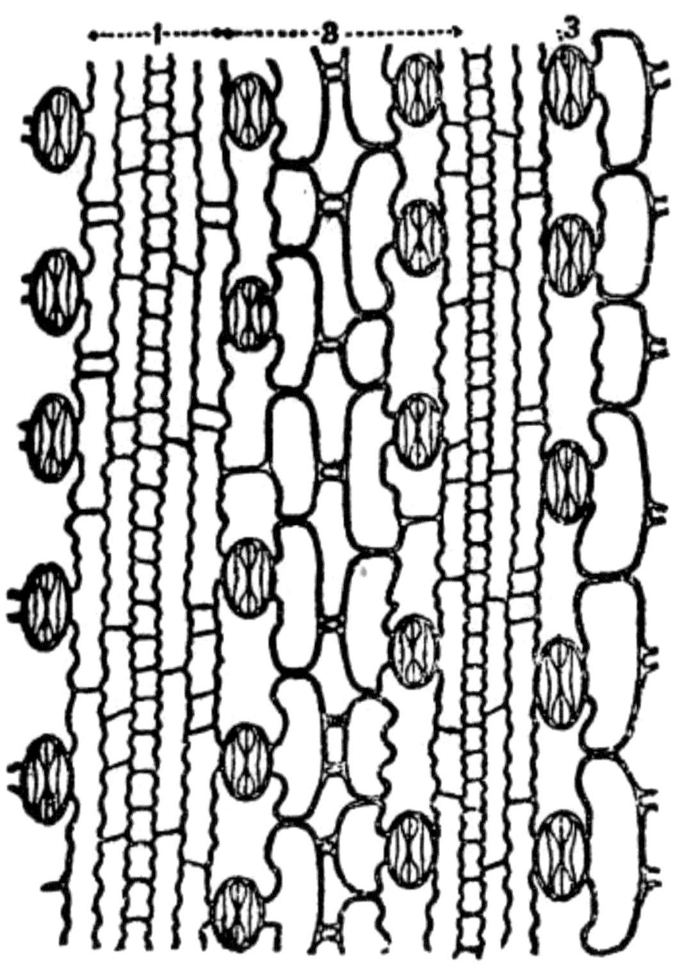

Fig. 62. — Upper epidermis of the leaf Panicum Isachne. × 200
1. Cells overlying the veins; 2. cells overlying the furrows; 3. stomata.

Fig. 63. — Lower epidermis of the leaf Panicum Isachne. × 200
1. Cells overlying the veins; 2. cells overlying the furrows; 3. stomata.

The epidermis covering the leaves consists of elongated cells with plane or sinuous walls, various kinds of short cells intercalated between the ends of long cells, motor-cells and stomata. Hairs of different sorts occur as outgrowths of the epidermis. The roughness of the surface of the leaves of grasses is due to the presence of very minute short hairs borne by the epidermis. In most cases these short hairs are found in regular rows. Although the epidermis is more or less even in the leaves of several grasses such as *Panicum repens, P. flavidum* and *Eriochloa polystachya*, it is wavy or undulating in the leaves of a few grasses. For example, the upper epidermis in the leaves of *Panicum fluitans* is undulating as it follows the contour of the ridges and furrows.

The epidermal cells have even surfaces in the leaves of most grasses but in some they bulge out. In the leaves of *Panicum flavidum* the cells of the lower epidermis are quite even, whilst those of the upper epidermis bulge out. The cells of both the upper and the lower epidermis are distinctly bulging out in the [Pg 40] leaves of *Panicum colonum*. In *Panicum fluitans* the cells of the upper epidermis bulge out so much as to form distinct papillæ.

The free surface of the epidermis is more or less cutinised in the leaves of all grasses. In some leaves the cuticle is very thick and even papillate as in the leaves of *Aristida setacea* and *Panicum repens* whilst in others it is very thin, as in the leaves of *Panicum colonum* and *P. fluitans*. Cutinisation is rather prominent in the leaves of grasses growing under dry conditions and it is less pronounced in mesophytic grasses.

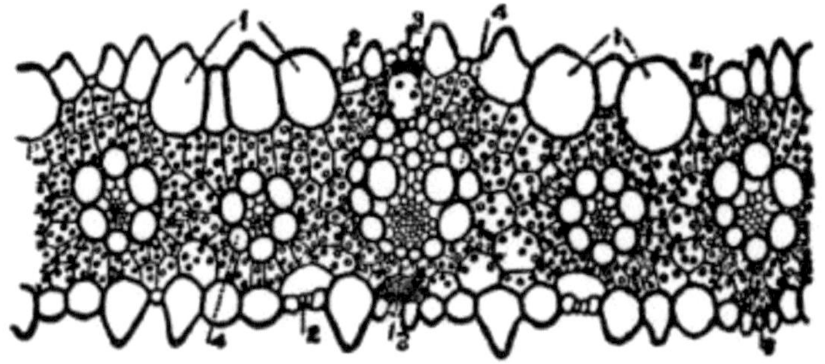

Fig. 64.—Transverse section of a portion of the leaf of Panicum colonum. × 120
1. Motor cells; 2. stomata; 3. sclerenchyma; 4. chlorophyllous layer.

Fig. 65.—Transverse section of a portion of the leaf of Panicum fluitans. × 120

1. Motor cells; 2. stomata; 3. sclerenchyma; 4. chlorophyllous layer.

As regards size, the epidermal cells overlying the sclerenchyma
are small and those lying over parenchyma are larger. Amongst the
larger cells some may be motor-cells. The stomata occur in regular
rows between the vascular bundles and they are quite characteristic
of grasses. They are more or less similar in structure in all grasses.
In the leaves of many grasses stomata are found in both the upper
and the lower epidermis and they are confined to the lower epidermis in a few grasses only.

The motor-cells vary very much both as regards their shape and
position. In some leaves as in the leaves of the grass *Panicum flavidum* the motor-cells are confined to the midrib on the upper surface.

[Pg 41]

The epidermal cells of this leaf are large and uniformly round.
(See figs. 66 and 67.)

Fig. 66.—Transverse section of a leaf of Panicum flavidum. × 20
1. Vascular bundle; 2. sclerenchyma; 3. motor-cells.

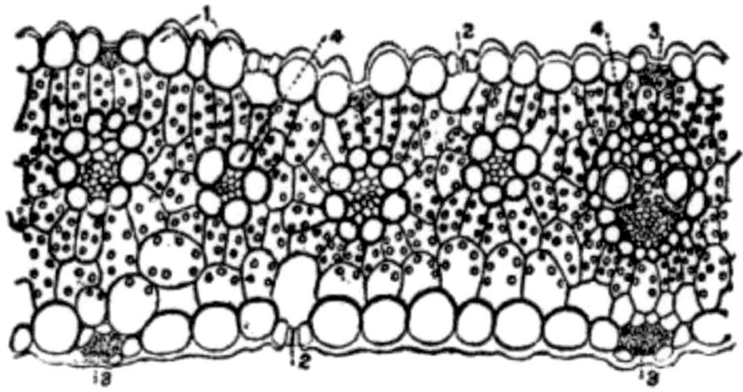

Fig. 67.—Transverse section of the leaf of Panicum flavidum. × 150
1. Motor-cells; 2. stomata; 3. sclerenchyma; 4. chlorophyllous layer.

In the case of most grasses the motor-cells are found in groups of three, four or five between the vascular bundles. The central motor-cell is usually the largest and it is somewhat obovate in shape in a transverse section of the leaf. In the leaves of *Panicum javanicum* and *Eriochloa polystachya* there are three or four motor cells in the group and the group consists of four, five or rarely six motor cells in the leaves of *Eragrostis Willldenoviana*. When there are distinct furrows between ridges these cells lie in the furrows and they are many in number. In the leaves of *Panicum repens* there are five to seven motor-cells in the furrows and the single row of cells stretched between the motor-cells and the lower epidermis in the furrow consists of more or less clear cells with sparsely scattered small chlorophyll grains. (See fig. 52.) The motor-cells occupying the furrows in the leaves of *Aristida setacea* are more in number than in *Panicum repens* and are of a different shape. All the cells lying in the furrow between the motor-cells and the sclerenchyma are clear cells free from chlorophyll grains.

Although the motor-cells differ in shape from the ordinary epidermal cells in most grasses, there are, however, a few grasses in which the motor-cells do not differ very much from the epidermal cells except in size. For example, in the leaves of *Panicum colonum* the motor-cells are just like the ordinary epidermal cells in shape but are larger. (See fig. 64.)

[Pg 42]

Motor-cells are usually confined to the upper epidermis, but they may also be found in the lower epidermis. In the leaves of *Pennisetum cenchroides* motor-cells are found in both the upper and the lower epidermis, the group in the upper epidermis alternating with that in the lower.

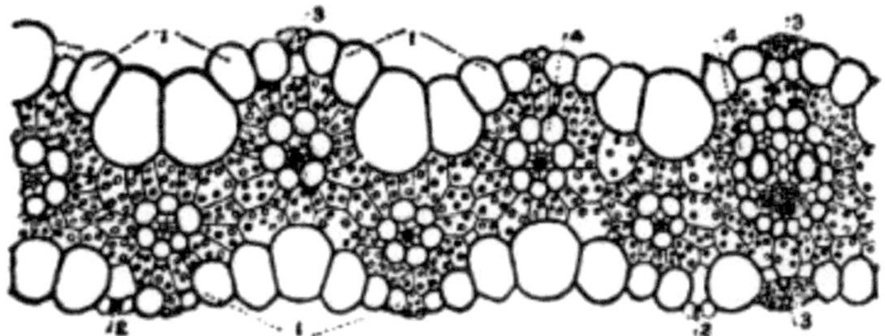

Fig. 68.—Transverse section of a portion of the leaf of Pennisetum cenchroides. × 100

1. Motor-cells; 2. stomata; 3. sclerenchyma; 4. chlorophyllous layer.

[Pg 43]

CHAPTER V.
CLASSIFICATION.

The family Gramineæ is usually divided into two series taking into consideration the presence or absence of a joint in the pedicel or rachis, the number of flowers in the spikelet and the position of the fertile flower. All the species in which there is a joint just below the spikelet, in the pedicel, in the rachis, or at the base of a cluster of spikelets come under one series **Panicaceæ**. The spikelets of the grasses coming under this series, when mature, fall away singly by themselves, or with their pedicels, or in groups with portions of the rachis. The spikelets are all similar and consist of usually four glumes. Each spikelet contains a single perfect flower and sometimes in addition a staminate flower just below the perfect flower. In this series the tendency for imperfection is always confined to the lower flowers, the terminal flower alone being perfect. For inclusion under this series the grass plant should have both the characters, articulation and position of the flower as mentioned above.

The second series **Poaceæ** includes those grasses in which the spikelets are one to many-flowered and continuous with their pedicels. But the rachilla of the spikelet may be jointed just above the empty glumes or between the flowering glumes. The complete flower is the lowest and the tendency for imperfection is in the upper flowers.

Of the two series the Panicaceæ appears to be more highly developed than the Poaceæ.

KEY TO TRIBES.

Series I.—Panicaceæ.

- A. Rachis of inflorescence not jointed.
    - o Spikelets 2-flowered; upper flower bisexual and lower male or neuter; the first glume the smallest I. Paniceæ.
    - o Spikelets 1-flowered;

- Spikelets articulate on their pedicels and falling away from them; flowers bisexual and usually with six stamens II. Oryzeæ.
  - Spikelets falling away with their pedicels; flowers bisexual or rarely imperfect III. Zoysieæ. [Pg 44]
- B. Rachis of inflorescence usually jointed.
  - Spikelets usually binate (3-nate at the top), pairs of spikelets alike or dissimilar; empty glumes larger and the flowering glumes smaller, hyaline, the fourth glume awned or reduced to an awn IV. Andropogoneæ.

Series II. — Poaceæ.

- A. Rachilla produced or not beyond the flowering glume.
  - Spikelets 1-flowered, with three glumes; first and second empty, third flowering and awned; rachilla jointed V. Agrostideæ.
  - Spikelets 1- or more-flowered, biseriate and secund on an inarticulate spike or on the spiciform branches of a slender panicle; flowers all or the lower only bisexual VI. Chlorideæ.
- B. Rachilla produced beyond the uppermost flowering glume and articulate.
  - Spikelets 2- or more-flowered, pedicelled, rarely sessile, in effuse, contracted or rarely spiciform panicles VII. Festucaceæ.
  - Spikelets 1- or more-flowered, sessile, 1- to 2- or more-seriate on the rachis of a simple spike, or partially sunk in cavities of the same. Glumes awned or not, first and second glumes are opposite or subcollateral, persistent or separately deciduous; first glume minute or absent VIII. Hordeæ.

[Pg 45]

CHAPTER VI.
Series I—Panicaceæ.
TRIBE I—PANICEÆ.

This is a fairly large and important tribe flourishing mostly in the warm regions and the tropics. It is very well represented in South India and fifteen genera are met with.

The inflorescence varies very much within this tribe and consists of spikes, racemes and panicles. The spikelets are usually four-glumed and contain one terminal perfect flower and a staminate or neutral flower below. But in the genus Isachne both the flowers are perfect. In some grasses the spikelets contain only staminate or pistillate flowers. In Coix and Polytoca the plant bears both male and female spikelets in the same inflorescence, but in Zea on the same plant they occur as distinct inflorescences. The littoral grass Spinifex is diœcious.

The first glume of the spikelet is the smallest. In Panicum it is nearly two-thirds or less than the third glume. It is very small in Digitaria and entirely suppressed in Paspalum. In Eriochloa it is reduced to a minute ridge lying just close to the swollen ring-like joint of the rachilla. The second and the third glumes are more or less equal and similar in texture. The fourth glume becomes firm and rigid along with its palea and usually encloses the grain.

The pedicel is jointed in some genera and in others it is continuous with the spikelet and not jointed. When mature the spikelets fall away either by themselves, singly with their pedicels or in groups with portions of rachis, according to the position of the joint. Bristles (branchlets) are often found on the pedicels. In Setaria a few are borne by the pedicels. The bristles form a regular involucre at the base of a group of spikelets in Pennisetum, and in Cenchrus these become united at the base into a mass forming a kind of burr around the spikelets.

KEY TO THE GENERA.

- A. Spikelets articulate on their pedicels.
    - o B. Spikelets without involucels.

- - C. Spikelets dorsally flattened, awnless.
    - Inflorescence racemed; glumes three; nerves of second glume five or less, side nerves curved 1. Paspalum.
    - Inflorescence digitate; glumes three with a minute glume; nerves of second glume five to seven, straight and prominent 2. Digitaria. [Pg 46]
    - Inflorescence panicled; glumes three with a thickening at the base of the spikelet 3. Eriochloa.
    - Inflorescence racemed or paniculate; glumes four, first two glumes unequal 4. Panicum.
    - Inflorescence panicled, branches of panicle produced beyond the uppermost spikelet; glumes four, the first being minute and hyaline 5. Chamæraphis.
    - Spikelets unisexual and diœcious 6. Spinifex.
  - CC. Spikelets awned.
    - Glumes four, second glume broadly fimbriate with hairs; palea of the third glume short and deeply cleft, fourth glume awned 7. Axonopus.
  - BB. Spikelets involucellate 8. Setaria.
- AA. Spikelets not jointed but continuous.
  - Spikelets in involucelled deciduous fascicles.
    - Involucre of bristles free 9. Pennisetum.
    - Involucre of bristles united 10. Cenchrus.

[Pg 47]

1. Paspalum, *L.*

These are annuals or perennials. The spikelets are plano-convex, orbicular to oblong, obtuse, secund, 2-ranked on the flattened or triquetrous rachis of the spike-like branches of a raceme, one-flowered and falling off entire from the very short or obscure pedicels. There are three glumes, all more or less equal and similar. The first and the second glumes are membranous, alike and as long as the third, the second glume is usually epaleate and occasionally with a minute palea. The third glume is chartaceous to sub-coriaceous and paleate. Lodicules are two and small. Stamens are three. The styles are slender and distinct with plumose stigmas exserted at the top of the spikelet. Grain is tightly enclosed in the third glume and its palea.

[Pg 48]

Fig. 69.—Paspalum scrobiculatum.

[Pg 49]

Paspalum scrobiculatum, *L.*

This is an annual grass, with stems tufted on very short rhizomes, erect or very shortly bent at base, glabrous, bifariously leafy and varying in height from 1 to 3 feet or more.

Leaf-sheaths are compressed, glabrous, loose, keeled, mouth hairy or not. The *ligule* is a short thin membrane. The *nodes* are glabrous.

The *leaf-blade* is linear-lanceolate, finely acuminate, keeled with a distinct midrib, and with very minutely serrulate margins, 6 to 18 inches by 1/12 to 1/3 inch.

The *inflorescence* consists of 2 to 5 sessile alternate spikes, usually distant and spreading and varying in length from 1 to 8 inches; the rachis is flattened and winged.

The *spikelets* are either orbicular or ovate-oblong, as broad as the rachis, glabrous, closely imbricating in two rows (rarely in three or four rows), sessile or rarely geminate on a common pedicel.

There are three glumes. The *first glume* is concave, 3- to 5-nerved (rarely 3- to 7-nerved). The *second glume* is flat, 5-nerved, with two strong sub-marginal nerves, sometimes with shallow transverse pits along the margins. The *third glume* is thickly coriaceous, brownish, shining, minutely striolate, margins roundly incurved throughout its length, paleate; the *palea* is similar to the glume in structure and colour, margins strongly inflexed and with two broad membranous auricles almost overlapping just below the middle. There are three *stamens*. The *stigmas* are white both when young and while fading. The style branches are diverging widely and then straight. There are two oblong cuneate fleshy *lodicules*.

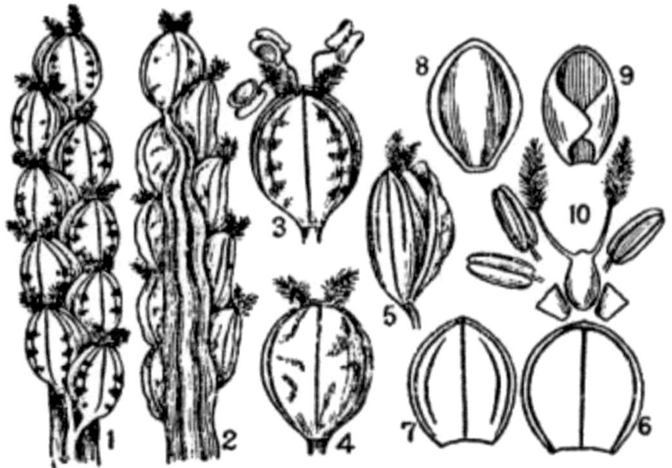

Fig. 70.—Paspalum scrobiculatum.
1 and 2. Front and back view of a portion of spike; 3, 4 and 5. spikelets; 6, 7 and 8. the first, second, and the third glume, respectively; 9.
palea of the third glume; 10. the ovary, stamens and the lodicules.

This grass flourishes all over the Presidency in moist places, such
as, bunds of wet lands, edges of ponds and lakes and in marshy
land. There are two forms of this grass, one with round [Pg 50] and
another with ovate oblong spikelets. They also vary in the size of
the spikelets—some forms have small spikelets and others large.
Sometimes the spikelets show variation in the number of glumes.

This grass is also cultivated for its grain. In cultivated forms the
spikelets are larger and the whole plant grows bigger. It is grown
both in wet and dry land.

Distribution.—Throughout India (wild and also cultivated).

[Pg 51]

2. Digitaria, *Rich.*

Annuals or perennials. The spikelets are lanceolate, 2- to 3-nate, in
digitate or racemose spikes, jointed on the pedicels but not thickened at the base, 1-flowered. There are usually four dissimilar
glumes in the spikelet. The first glume is hyaline very minute,

sometimes absent in the same species. The second glume is membranous, 1- to 5-nerved or nerveless. The third glume is membranous, almost equal to the fourth, usually 7- to 9-nerved, the nerves being straight, close, parallel and prominent, with a minute palea or without a palea. The fourth glume is chartaceous or sub-chartaceous, usually 3-nerved and paleate; palea is equal to and similar to the fourth glume, 2-nerved. Lodicules are two, small, broadly cuneate. Stamens are three. Styles are distinct with plumose stigmas exserted laterally near the apex of the spikelet. Grain is enclosed in the fourth glume and its palea.

KEY TO THE SPECIES.

- Spikelets 1/10 inch or more.
    - Spikes usually few, spikelets bearded 1. D. sanguinalis. Var. ciliaris.

    - Spikelets not bearded 2. Do. Var. extensum.

    - Spikes usually many; spikelets spreading 3. Do. Var. Griffithii.

- Spikelets less than 1/10 inch.
    - Spikes narrowly winged; spikelets subsilky with slender (not clavellate) hairs 4. D. longiflora.

[Pg 52]

Fig. 71.—Digitaria sanguinalis, Var. ciliaris.

[Pg 53]

Digitaria sanguinalis, *Scop.*

Var. ciliaris.

This is an annual grass either with erect tall stems or long prostrate stems, varying in length from 1 to 3 feet or more.

The *leaf-sheath* is herbaceous, loose and glabrous. The *ligule* is a distinct membrane. The *nodes* are glabrous.

The *leaf-blade* is linear-lanceolate or linear, flat, glabrous or very sparsely hairy, varying in length from 2 to 5 or 6 inches and in breadth from 1/6 to 1/3 inch.

The *spikes* are usually few, 2 to 6, 3 to 6 inches long, with a triquetrous, narrowly winged rachis.

The *spikelets* are oblong, acute, binate, one pedicel being shorter than the other, usually appressed to the rachis and not spreading.

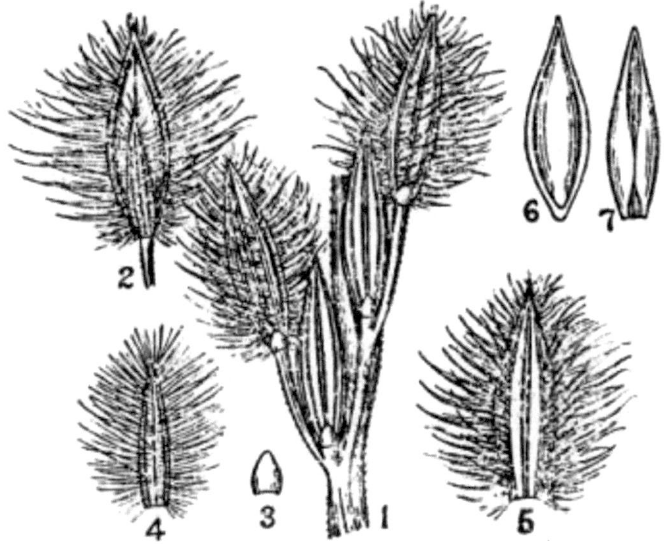

Fig. 72. — Digitaria sanguinalis. Var. ciliaris.
1. A portion of the spike showing the binate spikelets; 2. a spikelet; 3. the minute scale-like first glume; 4, 5 and 6. the second, third and the fourth glume, respectively; 7. the palea of the fourth glume.

There are four *glumes* including the minute glume. The *first glume* is a very minute scale. The *second glume* is about half as long as the third glume, membranous, usually 3-nerved and sometimes 3- to 5-nerved, distinctly ciliate. The *third glume* is oblong-lanceolate, acute, membranous, 3- to 5-nerved, sparingly hairy in the lower spikelet and densely bearded with soft spreading hairs in the upper spikelet.

The *fourth glume* is lanceolate, or oblong-lanceolate, acute, some-what chartaceous, paleate; *palea* is like the glume in texture. *Anthers* are pale yellow. *Stigmas* are white. There are two small cuneate *lodicules*.

This is an excellent fodder grass. It grows well in all kinds of soils, rich or poor, and is very common in dry fields brought under culti-vation.

Distribution. — Throughout India.

[Pg 54]

Digitaria sanguinalis, *Scop.*

Var. Griffithii.

This is an annual with stems ascending from a prostrate or genic-ulate base, glabrous and varying in length from 1 to 3 feet.

The *leaf-sheath* is glabrous, thinly herbaceous and loose. The *ligule* is a distinct membrane and the *nodes* are glabrous.

The *leaf-blade* is linear or linear-lanceolate, flat, acuminate, varying in length from 2 inches to 12 inches and in breadth 1/6 to 1/3 inch.

The *inflorescence* is of several slender spikes, usually drooping, 2 to 4 inches; the rachis is filiform and trigonous.

The *spikelets* are linear-lanceolate, solitary or in distant pairs, gla-brous or ciliate, pedicelled and when binate the upper pedicel often longer than the spikelets, usually spreading and not appressed to the rachis.

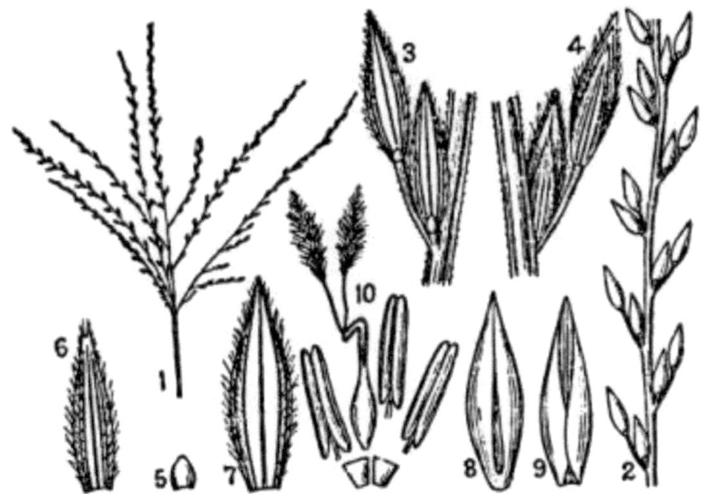

Fig. 73.—Digitaria sanguinalis, Var. Griffithii.
1. Inflorescence; 2. a portion of the spike; 3 and 4. sessile and pedicelled spikelets front and back view, respectively; 5. the scale-like
first glume; 6, 7 and 8. the second, third and the fourth glume, respectively; 9. palea of the fourth glume; 10. the lodicules, stamens
and the ovary.

There are four *glumes*. The *first glume* is a minute scale. The *second glume* is shorter than the third and narrower, 5-nerved, ciliate, acute or sometimes with two fine teeth. The *third glume* is oblong-lanceolate, acute, 5-nerved (rarely 3-nerved), ciliate on the nerves. The *fourth glume* is lanceolate, acute, sub-chartaceous, paleate; *palea* is like the glume in texture. *Anthers* are yellow and *stigmas* are white. *Lodicules* are two and small.

This seems to be a good fodder grass. It grows in all kinds of soils. It is not so common in the plains as on the hills, though it occurs in the plains at the base of the hills.

Distribution.—Throughout India.

[Pg 55]

Fig. 74. — Digitaria sanguinalis, Var. extensum.

[Pg 56]

Digitaria sanguinalis, *Scop.*

Var. extensum.

This grass is an annual with stems ascending from a prostrate or geniculate, rooting branched base, greenish or purplish, glabrous and varying in length from 1 to 2-1/2 feet.

The *leaf-sheath* is thin, herbaceous, rather loose, keeled and glabrous. The *ligule* is a distinct membrane, truncate, rarely irregularly toothed. The *nodes* are glabrous.

The *leaf-blade* is linear-lanceolate, acuminate, flat when mature and convolute when young, glabrous, 1 to 12 inches long and 1/6 to 1/3 inch broad, the margin is very closely and finely serrate, the midrib is prominent with three or four main veins on each side.

The *inflorescence* consists of a few or many spikes, corymbosely arranged on a short angular slightly rough axis, erect or spreading, 1-1/2 to 4 inches long, the lowest ones in whorls of two to four; the rachis is nearly triquetrous, laterally winged, base thickened and with a few long white hairs; the peduncle is cylindric, smooth, 6 to 12 inches long.

Fig. 75. — Digitaria sanguinalis, Var. extensum.
1. A portion of spike; 2, 3 and 3a. the back and front views of a

spikelet; 4, 5 and 6 the first, second and the third glume, respectively; 7. palea of the third glume; 8. anthers, lodicules and the ovary.

The *spikelets* are oblong-lanceolate, acute, about 1/10 inch long, binate, one pedicelled and the other subsessile, the pedicel is angular, about 1/2 to 2/3 the length of the spikelet.

There are three *glumes* in the spikelet corresponding to the second, third and fourth glumes of a Panicum, the first glume being obsolete. The *first glume* is membranous, ovate-lanceolate, acute, about 1/3 the length of the spikelet or very much less, 3-nerved, densely ciliate along the margins and silkily hairy between the nerves. The *second glume* is greenish, oblong lanceolate, acute, [Pg 57] ciliate along the margins and with fine appressed silky hairs between the lateral nerves, 5-nerved, palea is very minute or absent. The *third glume* is oblong, sub-acuminate, a little shorter than the second glume, 3-nerved, sub-chartaceous, paleate; *palea* is similar to the glume in texture. *Anthers* are pale yellow with a tinge of purple. *Stigmas* are white. *Lodicules* are two, minute and cuneate.

This is an excellent fodder grass and is very much liked by cattle. It grows very rapidly and is found in cultivated fields and in somewhat rich loamy soils.

Distribution. — Throughout the Presidency in the plains and low hills.

[Pg 58]

Fig. 76.—Digitaria longiflora.

[Pg 59]

Digitaria longiflora, *Pers.*

This is a perennial grass with short underground branches covered with scales. Stems are many, tufted, slender, creeping and rooting, or ascending and suberect, simple or branched, 6 to 20 inches long and leafy and leaves bifarious and divaricate.

Leaf-sheaths are hairy or glabrous, compressed, keeled. The *ligule* is a short membrane. *Nodes* are glabrous.

Leaf-blades are broadly lanceolate or linear-lanceolate, acute, spreading, flat, or in short-leaved forms, stiff and pungent, 1 to 2 inches long (rarely also 5 inches long), glabrous above and below, ciliate at the margins towards the base, and with a very minutely serrate hyaline margin.

The *inflorescence* consists of two to four terminal spikes with a slender, long, hairy or glabrous peduncle. The spikes are slender, erect or spreading with fine winged glabrous rachis.

The *spikelets* are small, 1/20 to 1/14 inch, geminate, one short and the other long pedicelled, appressed to the rachis, elliptic, silky with slender crisped hairs, pale green or purplish.

Fig. 77.—Digitaria longiflora.
1. A portion of the spike; 2. the first glume; 3 and 4. the second and third glumes; 5 and 6. the fourth glume and its palea; 7. lodicules, ovary and stamens.

There are three *glumes* with a rudimentary first glume. The *first glume* is very minute and hyaline. The *second glume* is as long as the third, membranous, 5-nerved (rarely 3- to 7-nerved), silkily hairy.

114

The *third glume* is similar to the second and usually 7-nerved (rarely 3- to 5-nerved). The *fourth glume* is sub-chartaceous, ovate-oblong, paleate, slightly shorter than the third glume, pale brown, smooth. There are two small *lodicules*. Styles are long and purple.

This grass grows in cultivated dry fields. It seems to like a sandy loamy soil.

Distribution. — Throughout India.

[Pg 60]

3. Eriochloa, *H. B. & K.*

These are annuals or perennials. Leaves are flat. The inflorescence is a raceme or a panicle. Spikelets are one-flowered, borne unilaterally on the branches, and the base is thickened and jointed on the top of a short pedicel. The spikelet has three glumes. The first and the second glumes are subequal, membranous. The third glume is apiculate, hardened in fruit. The lodicules are small and truncate. There are three stamens with linear anthers. Styles are two free, with plumose stigmas. The grain is oblong, free within the hardened glume and its palea.

[Pg 61]

Fig. 78.—Eriochloa polystachya.

[Pg 62]

Eriochloa polystachya, *H. B. & K.*

This grass is a densely tufted perennial, varying in height from 2 to 3 feet, with a short creeping root-stock. Stems are slender, or stout, simple and branching, ascending from a short creeping and rooting base, glabrous, slightly channelled on one side.

The *leaf-sheath* is glabrous, green or partly purplish, striate, loose, mouth and margins above sometimes pubescent. The *ligule* is a short villous ridge. *Nodes* are perfectly glabrous.

The *leaf-blade* is flat, linear or linear-lanceolate, acuminate, glabrous on both sides, with a slender or prominent midrib, veins more or less uniform, 2 to 10 inches long and 1/6 to 1/3 inch wide, convolute when young. Sometimes the blade is purplish below.

The *inflorescence* is a panicle on a long or short glabrous stalk, striate, 2 to 7 inches long, with four to fifteen erect or spreading, lax branches, the main rachis is glabrous, angular and deeply grooved. Spikes or branches are slender, alternate, 1 to 2-1/2 inches, becoming shorter upwards, thickened and puberulous at the base, and the secondary rachis is flexuous, grooved, angular, and obscurely pubescent.

Fig. 79.—Eriochloa polystachya.

1. A portion of the branch; 2, 3 and 4. the first, second and the third glume, respectively; 4a. back view of the third glume; 5. palea of the third glume; 6. lodicules, stamens and the ovary; 7. grain.

The *spikelets* are green or purplish, ovate, lanceolate, acuminate 1/8 to 1/6 inch long, softly hairy, stalked, solitary above and binate below and then one with a long and the other with a short pedicel rising from a common short branchlet, loosely imbricate, distichous and shortly stipitate and the stipe with a purple thickening; pedicel is short, 1/24 to 1/12 inch with sometimes long deciduous hairs and the tip somewhat thickened.

There are three *glumes* in the spikelet. The *first glume* is membranous, covered densely with silky hairs, ovate-lanceolate, acuminate, tip very minutely 3-toothed with three to five fine nerves. The *second glume* is similar to the first glume but with a more pointed tip, faintly 3- to 5-nerved; *palea* is not present and if present [Pg 63] it is very small, hyaline and empty. The *third glume* is shorter than the first and the second glumes, thinly coriaceous, punctate, oblong, obtuse, pale, faintly 3- to 5-nerved with a short scaberulous awn, paleate; *palea* is oblong, similar to the glume in texture, margin infolded. *Anthers* are three, linear, pale yellow. *Stigmas* are feathery, white when young and purple later. *Lodicules* are two and distinct.

This is a common succulent grass growing in large or small tufts in moist situations such as sides of water channels, rivulets and bunds of paddy fields. It is very much liked by cattle. This grass is easily recognized by the silky lanceolate spikelets which have a purple thickening at the base.

Distribution.—Plains of India and Ceylon and in all hot countries.

[Pg 64]

4. Panicum, *L.*

The grasses of this genus are annual or perennial and of various habits. Inflorescence is either a raceme of spikes or, a lax or contracted panicle. Spikelets are small, solitary or two to four, rarely more ranked, 1- to 2-flowered, ovoid or oblong, rounded, or dorsally or laterally compressed, falling entire with the pedicels. There are four glumes in a spikelet. The first two glumes are empty and the first glume is small (sometimes minute) and fewest nerved. The second glume is equal or very nearly equal to the third glume, oblong-ovate or lanceolate, 5- to many-nerved. The third glume is similar to the second, male or neuter, paleate or not, 3- to 9-nerved. The fourth glume is chartaceous, sometimes shortly stalked, ovate-oblong or lanceolate, hardened in the fruit, smooth or rough, bisexual, paleate; the palea is as long and of the same texture as the glume. Lodicules are cuneate or quadrate and two in number. There are three stamens and an ovary with two style branches ending in feathery stigmas. Grain is free and enclosed by the hardened fourth glume and its palea.

KEY TO THE SPECIES.

- A. Inflorescence racemose of simple (rarely branched) spikes bearing secund spikelets.
    - I. Rachis of spikes broad and flattened.
        - (a) Spikelets biseriate.
            - Spikelets villous. 1. P. Isachne.
            - Spikelets glabrous.
                - Spikes shorter than the internodes. 2. P. flavidum.
                - Spikes longer than the internodes. 3. P. fluitans.
        - (b) Spikelets 3- to 5-seriate.
            - Third glume awned.
                - Stems stout, erect. 4. P. Crus-galli.
                - Stems stout, prostrate at base. 5. P. stagninum.
            - Third glume cuspidate.
                - Stems slender. 6. P. colonum.
    - II. Rachis of spikes narrow, filiform, terete or angular.
        - First glume shorter than the third.

- - First glume semilunate, about 1/4 of the third glume. 7. P. prostratum.
    - First glume 1/2 of or less than 1/2 of third glume, 5-nerved. 8. P. ramosum.
  - Leaf base broad or cordate.
    - Fourth glume shortly awned. 9. P. javanicum.
    - Fourth glume muticous. 10. P. distachyum.
- B. Inflorescence a contracted or open panicle.
  - I. Panicle contracted and spike-like.
    - Spikelets lanceolate and first glume minute. 11. P. interruptum.
  - II. Panicle effuse.
    - Annuals; first glume nearly 3/4 of the third glume. 12. P. trypheron.
    - Perennials; first glume less than 1/3 of the third glume. 13. P. repens.

[Pg 65]

120

Fig. 80.—Panicum Isachne.

[Pg 66]

Panicum Isachne, *Roth.*

This is an annual grass usually growing in tufts with fine fibrous roots and many slender spreading branches, all of them at first creeping and horizontal, rooting at the nodes and then becoming erect and varying in length from 1 to 2 feet.

Stems are very slender, glabrous or covered with scattered hairs, purplish or pale green, and branching freely towards the base.

The *leaf-sheath* is shorter than the internodes, green or purplish, striate, externally hairy with scattered bulbous-based hairs, varying in length from 1/2 to 3 inches, the outer margin of the sheath is ciliate with long hairs and at the mouths sometimes long hairs are present, especially when the leaves are young. The *ligule* is merely a dense fringe of long hairs. *Nodes* are tumid, purplish, covered with long hairs.

The *leaf-blade* is flat but convolute when young, lanceolate or linear-lanceolate, acuminate, base rounded and margin with minute serrations. It is glabrous or occasionally hairy with scattered, tubercle-based, deciduous hairs, and varying in length from 1 to 3 inches generally (sometimes in well-grown plants it is 5 inches) and in breadth from 1/8 to 1/4 inch. The midrib is prominent though slender at the base and four veins are present on each side with five or six smaller ones between them.

Fig. 81.—Panicum Isachne.
1 and 1a. Front and back view of a spike; 2 and 2a. back and front

views of a spikelet; 3 and 4. the first and the second glume, respectively; 5 and 5a. the third glume and its palea; 6 and 6a. the fourth glume and its palea; 7. lodicules, anthers and ovary; 8. grain.

The *inflorescence* is an erect, narrow panicle consisting of spikes varying in number from 5 to 12 and in length from 2 to 3 inches. The *spikes* are erect, pressed to the very slender rachis, longer than the internodes of the main rachis, stalked or sessile, mostly simple but sometimes the lower dividing into two or three branches, 1/2 to [Pg 67] 1 inch long. The rachis of the spike is very slender, angular, flexuous, narrower than the spikelets, scaberulous with a few long cilia at the angles.

The *spikelets* are very small, 1/16 inch long, turned all to one side and closely packed in two rows, oblong or oval-oblong, obtuse or subacute, softly hairy, pale green or purplish, with very short pedicels which are pubescent with a few long hairs towards the thickened cupular tips.

There are four *glumes* in the spikelet. The *first glume* is very small, membranous, glabrous, broader than long, cordate or triangular, broadly but shallowly emarginate, nerveless or very obscurely 1- to 2-nerved. The *second glume* is pale or purplish, 5-nerved, hairy, as long as the third glume, membranous, oblong and obtuse. The *third glume* is pale, nearly equal to the second glume with a longitudinal depression at the back, less hairy than the second glume, 3-nerved (rarely 5-nerved also); *palea* is present, and it is hyaline, shorter than the glume, truncate or shallowly retuse, usually barren but occasionally with three stamens. The *fourth glume* is oblong, rounded, coriaceous, smooth, shining, dorsally flattened, 3- or indistinctly 5-nerved; *palea* is similar to the glume in texture and with folded margins. There are three *stamens* with yellow anthers. *Lodicules* are two, very small and distinct. *Ovary* has two styles with feathery *stigmas* white at first, but turning deep purple while withering.

This delicate and small grass occurs here and there as mere tufts especially in sheltered situations. It usually flourishes in black cotton soils amidst cholam (*Andropogon Sorghum*), although it thrives equally well in other rich soils. This is considered to be a very good fodder grass.

Distribution.—It is fairly common all over the Madras Presidency, and goes up to 3,000 or 4,000 feet. It occurs in Africa, America and Italy.

[Pg 68]

Fig. 82.—Panicum flavidum.

Panicum flavidum, *Retz.*

This plant is a tufted annual. It branches freely from the base; branches are tufted, decumbent at first but soon becoming erect, slender, glabrous, compressed and leafy, varying in length from 1 to 3 feet.

Leaves are somewhat distichous. The *leaf-sheath* is compressed, glabrous, sometimes with a tinge of purple, the lower ones swollen at the base and the mouth is hairy. The *ligule* is a fringe of hairs. Nodes are glabrous.

The *leaf-blade* is flat, thinly coriaceous, linear-lanceolate and acuminate, or ligulate with a rounded tip, 3 to 5 inches in length, 3/16 to 5/16 inch wide, glabrous or very thinly scaberulous, base rounded or slightly cordate with long white ciliate hairs on the small basal lobes.

Fig. 83. — Panicum flavidum.
1 and 2. Front and back view of a portion of spike; 1a and 2a. the front and back view of a spikelet; 3 and 4. the first and the second glume, respectively; 5 and 5a. the third glume and its palea; 6 and 6a. the fourth glume and its palea; 7. anthers and ovary; 8. grain.

The *inflorescence* is a raceme of spikes, 5 to 10 inches long, erect or inclined on a short or long, glabrous, strongly channelled peduncle; the main rachis is grooved, angled and scaberulous. *Spikes* are few or many, 1/4 to 1 inch long, erect, pressing on the rachis of the inflorescence along the groove, distant and sessile; the lower spikes are very much shorter than the internodes, but the upper equal to or longer than the internodes; the rachis of the spike is angular, flattened below, erect or slightly recurved.

The *spikelets* are white, in two rows on a flattened rachis, obliquely ovoid or gibbously globose, glabrous, sessile, 1/8 inch in length.

There are four *glumes*. The *first glume* is suborbicular, about half the length of the third glume, usually 3-nerved. The *second glume* is broadly ovate, obtuse, concave, larger than the first glume and nearly equal to or shorter than the fourth glume, 7-nerved, rarely 7- to 9-nerved, nerves are anastomosing, tip rounded. The *third glume* is broadly ovate or oblong, equal to or longer than the [Pg 70] fourth glume, obtuse, 3- to 5-nerved, paleate, mostly with and rarely without stamens. The *anthers* are yellow and they do not open until the stigmas and anthers of the fourth glume are thrown out. *Lodicules* are two and conspicuous. *Palea* is hyaline with infolded margins. The *fourth glume* is coriaceous, broadly ovate, tip acutely pointed and almost cuspidate or acute, mucronate, white or brownish, reticulately minutely pitted. *Anthers* are three and yellow. *Stigmas* are purplish. *Lodicules* are small but conspicuous.

This grass is very common throughout the plains and grows in the bunds of paddy fields and in wet situations, and goes up to moderate elevations on the hills. Cattle eat this grass greedily and seem to like it. It is considered to be an excellent fodder.

Distribution. — In wet situations all over India ascending to 6,000 feet. Occurs also in Ceylon, Africa, Tropical Asia and Australia.

[Pg 71]

Fig. 84.—Panicum fluitans.

[Pg 72]

Panicum fluitans, *Retz.*

This is a perennial grass with prostrate branches which afterwards become erect towards the free ends. The young branches are covered with scale-leaves. Stems are stout, glabrous, smooth and hollow, rooting at the lower nodes.

The *leaf-sheath* is loose, glabrous, striate, margins not ciliate. The *ligule* is a ridge with a row of erect long hairs. *Nodes* are glabrous.

The *leaf-blade* is firm, linear, finely acuminate, base rounded, rather narrower than the sheath at the white band, very thinly scaberulous above and glabrous below, veins prominent above, 3 to 9 inches long, 1/4 to 7/16 inch broad; margins are slightly incurved and the midrib is conspicuous only at the lower portion of the blade. The scale-leaves persist at the base of the stems.

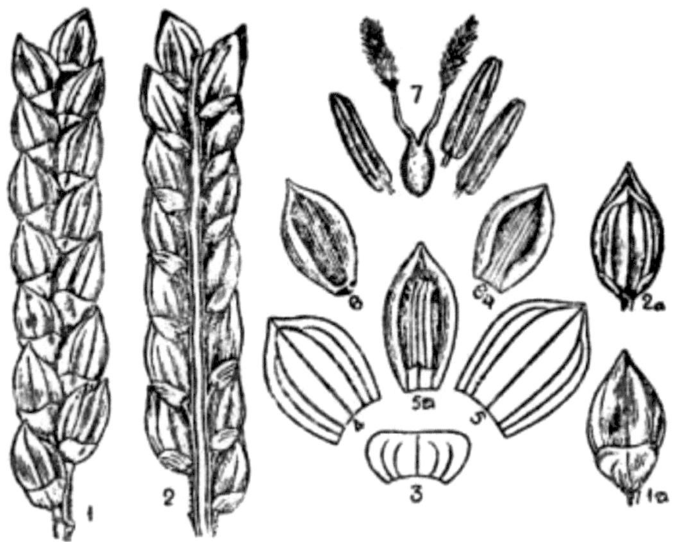

Fig. 85.—Panicum fluitans.
1 and 2. Front and back view of a spike; 1a. and 2a. front and back view of a spikelet; 3, 4 and 5. first, second and third glume respectively; 5a. palea of the third glume and stamens in it; 6 and 6a. fourth glume and its palea; 7. stamens and ovary.

The *inflorescence* is a compound spike varying in length from 4 to 10 inches, erect; the main rachis is triquetrous, dorsally rounded, glabrous and very thinly scaberulous at the edges. *Spikes* are many (fifteen and more), sessile, secund, generally longer than the internodes, and appressed to the rachis, 1/4 to 1-1/2 inches long; the rachis of the spike is angular, edges scaberulous and with very fine short hairs.

The *spikelets* are pale, ovoid, acute, biseriate, imbricate, very shortly pedicellate, glabrous, 1/16 to 1/8 inch, pedicels are hairy with a few long hairs towards the base.

There are four *glumes*. The *first glume* is white, thin, membranous, truncate and wavy at the apex, nerveless or sometimes with one to three short nerves, less than one-third of the third glume, broader than long and clasping at the base. The *second glume* is ovate, obtuse or subacute, concave, submembranous, slightly shorter [Pg 73] than the fourth glume, 5-nerved but occasionally 6- or 7-nerved. The *third glume* is a little longer than the second and the fourth, usually 5-nerved, broadly ovate, acute, paleate, always with three stamens which come out only after the fading of the stigmas and enlargement of the ovary in the fourth glume. *Lodicules* are distinct and conspicuous; *palea* is broad with incurved broad margins and hyaline. The *fourth glume* is thinly coriaceous, shining, striolate, broadly ovate, mucronate, compressed, faintly and thinly 5-nerved and *palea* with infolded margins. *Anthers* are yellow. *Stigmas* are white when young. *Lodicules* are distinct.

It is a common grass of the wet lands met with in many parts of the Presidency and often confused and united with *Panicum punctatum*, Burm.

Distribution.—Throughout India and Ceylon. It is also found in Arabia, Afghanistan, Africa and Tropical America.

[Pg 74]

Fig. 86. — Panicum Crus-galli.

[Pg 75]

Panicum Crus-galli, *L.*

It is a tufted annual with many erect branches growing to a height varying from 2 to 3 or 4 feet and the whole plant is glabrous. Stem is stout or slender, simple or branched.

The *leaf-sheath* is smooth, glabrous and loose, varying in length from 2 to 6 inches, keeled. The *ligule* is only a smooth semilunar line without hairs. *Nodes* are glabrous and the lower nodes bear adventitious roots.

The *leaf-blade* is narrowly linear-lanceolate, flat, finely acuminate, glabrous or very minutely scabrid with a stout midrib; margin is minutely serrate and with tubercle-based hairs near the base. The blades of the lower leaves are longer than those in the upper and at the junction with the sheath the blade is narrow, just as broad or less than the sheath, and becomes broader about the middle; the length varies from 6 to 10 inches generally, also to 14 inches, and breadth at base 1/4 inch and at the middle 5/16 inch; the upper leaf-blade is generally shorter, varying from 5 to 10 inches and very broad at the base near the sheath, about 7/16 inch and gets gradually narrow upwards. It is convolute when young.

The *inflorescence* is a compound spike varying in length from 4 to 8 inches, contracted and pyramidal and always erect; the main rachis is stout, angled with very minute hairs on the ridges and with a tuft of bristly hairs and also tubercle-based hairs at the place of insertion of the spikes. *Spikes* are many (up to 16 or rarely more), simple or branched, the lower ones longer, but getting gradually shorter upwards, and varying in length from 1/2 to 2 inches. The rachis of the spike is angular, with scattered tubercle-based bristly hairs.

Fig. 87.—Panicum Crus-galli.
1 and 2. Front and back views of spike; 3. spikelet; 4 and 5. first and second glumes; 6 and 7. third glume and its palea; 8. fourth glume, front and back view; 9. ovary, anthers and lodicules.

The *spikelets* are turgid, densely packed on one side of the rachis in three to five rows, sessile or subsessile, sub-globose or ovoid, [Pg 76] with unequal tubercle-based bristly hairs on the nerves of the glumes and with short minute hairs on the outer surface of the glumes, 1/12 to 1/8 inch; awn 1/4 inch to 5/16 inch.

There are four *glumes*. The *first glume* is 1/3 to 1/2 of the third glume, suborbicular, abruptly acuminate or rarely mucronate and 5-nerved (very rarely 5- to 7-nerved), clasping at base and margins thinly ciliolate. The *second glume* is ovate oblong, short, awned and 5-nerved; sometimes with partial nerves at the apex between the central and the lateral nerves, and then 5- to 7- or 5- to 9-nerved, hispidly hairy on the nerves, margins ciliolate. The *third glume* is as long as the second, ovate-oblong and the apex abruptly ending in a stout scabrid nerved awn, varying in length from 1/4 to 3/8 inch, rarely 1 inch; 5- to 7-nerved (two partial at tip), paleate and sometimes with three stamens; *palea* is hyaline, ovate-oblong with infolded margins. The *fourth glume* is smooth, shining, broadly oblong, faintly 5-nerved, apex rounded or cuspidate with a few cilia; paleate with a single bisexual flower; *palea* is similar to the glume in structure. *Anthers* are orange yellow, and *lodicules* are very small. *Stigmas* are white. Grain is smooth and ovoid.

This grass grows in paddy fields and wet places generally. It is considered to be a very good fodder grass in Australia and America. This is the "Barn-yard" grass of the Americans, highly valued as a fodder grass.

Distribution. — Throughout India in wet places and in paddy fields.

[Pg 77]

Panicum stagninum, *Retz.*

It is an annual. The stems are glabrous, creeping and somewhat prostrate at the base, and the upper portion is erect, 3 to 4 feet long, and rooting at the nodes in the geniculate portion of the stem.

The *leaf-sheath* is smooth, striate, glabrous, sometimes pubescent about the lower nodes, varying in length from 1-1/2 to 4-1/2 inches. The *ligule* is distinct, consisting of a fringe of stiff hairs.

The *leaf-blade* is linear-lanceolate, acuminate or acute, base rounded, glabrous, smooth below, especially in the lower part, and scabrid above and in the upper part, 6 to 12 inches long, by 1/4 to 3/8 inch; the lower leaves have their blades somewhat narrower at the base than in the middle, but the blades in the upper part of the stem and in the middle are of the same breadth; margins are very minutely serrate.

Fig. 88.—Panicum stagninum.
1. Front view of a portion of spike; 2. back view of the same; 3 and 4. front and back views of a spikelet; 5, 6 and 7. the first, second and the third glume, respectively; 8. palea of the third glume with its anthers; 9. front and back view of fourth glume; 10. the ovary, stamens and lodicules.

The *inflorescence* is 4 to 8 inches long; the main *rachis* is angular, grooved, scabrid on the ridges. The *spikes* are 7 to 10 inches, alternate, pale green or purplish, rather distant, spreading or suberect (never erect) 1/2 to 1-1/2 inches long, sessile and with a tuft of bristly hairs at the base; the rachis of the spike is angular, grooved with scattered bulbous-based bristles on the ridges.

The *spikelets* are four ranked, ovoid-lanceolate, 1/8 to 1/6 inch long without the awn, somewhat flattened on one side and gibbous on the other, pale green or purplish, with equal bulbous-based bristly hairs on the nerves.

There are four *glumes*. The *first glume* is half of the third glume, thin, membranous, hairy, broadly ovate, abruptly cuspidate at the apex, and acuminate, 5-nerved (rarely 3-nerved). The *second glume* is broadly ovate-lanceolate, concave, acuminate, short awned, 5-nerved with two partial nerves one on each side of the central [Pg 78] nerve (7- to 9-nerved at the tip), hairs on nerves, a few tubercled.

The *third glume* is similar to the second, broadly ovate-lanceolate, awned, awn 1/8 to 1/4 inch, paleate with usually three stamens, occasionally neuter. *Lodicules* are present. The *fourth glume* is chartaceous, shining, smooth ovate-oblong, apex cuspidate, with a few hairs on the edges at the apex, faintly 5-nerved. The *anthers* within this glume come out before those of the third glume. *Anthers* are three, yellowish and *lodicules* are conspicuous though small.

In this grass very often, purple streaks or bands occur across the leaf blades and the sheath and the spikelets become purple on one side as is met with in P. colonum. This grass is occasionally found in the paddy fields either alone, or along with *Panicum Crus-galli.*

Distribution. — Throughout the Madras and the Bombay Presidencies and in Ceylon in wet places especially in cultivated ground and in ditches. Occurs more or less throughout India.

[Pg 79]

Fig. 89.—Panicum colonum.

[Pg 80]

Panicum colonum, *L.*

This is a slender annual growing to a height of 2 feet. The stems are creeping below, erect above, and with roots in the lower internodes of the decumbent part of the stem, smooth, dull green or partly purplish.

The *leaf-sheath* is glabrous and sharply keeled. The *nodes* are glabrous or obscurely pubescent. There is no *ligule*.

The *leaf-blade* is narrow, lanceolate, acuminate, glabrous but sometimes tubercle-based hairs occur just on the margin at the base of the leaf-blade close to the white band, varying in length from 1 to 6 inches and in breadth 3/16 to 5/16 inch; the margin is minutely and distantly serrate, midrib is quite distinct and there are three main veins on each side and three or four smaller between main ones. The blades of the lower leaves are narrow at the base and broader at about the middle but those of the upper are equally broad at the base, as well as at the middle.

Fig. 90. — Panicum colonum.
1 and 2. The front and back view of the spikes; 1a and 2a. the back and the front view of the spikelet; 3, 4 and 5. the first, second and the third glume, respectively; 6. palea of the third glume; 7 and 8. the fourth glume and its palea; 9. ovary anther and lodicules.

The *inflorescence* is a contracted panicle, 3 to 5 inches long. *Spikes* are from 8 to 20, suberect, usually distant, 1/4 to 1 inch long and getting shorter upwards; the rachis of the spike is stout, angular, scaberulous on the angles with a few long hairs towards the base.

The *spikelets* are small green or partly purplish 1/12 to 1/10 inch long, globosely ovoid, acute, pubescent with minute hairs on the outer surface of the glumes and bristly hairs on the nerves, all on one side, sessile or very shortly pedicelled, two or three from a node, one or two barren, 3- to 5-seriate.

There are four *glumes* in the spikelet. The *first glume* is about half of the third glume, broadly ovate or suborbicular, acute, generally 3-nerved, rarely 5-nerved, pubescent between and hispidly hairy on the nerves. The *second glume* is as long as the third, broadly ovate, cuspidate, 5-nerved sometimes with two partial [Pg 81] nerves added one on each side of the central vein, pubescent between the veins and hispid on the veins. The *third glume* is similar to the second, 5-nerved, tip with a few cilia, paleate; *palea* is empty oblong-orbicular, subacute. The *fourth glume* is coriaceous, shining, turgid, broadly ovate, acute, paleate. Sometimes the tip possesses a few cilia. *Anthers* are three, pale yellow and *stigmas* dark purple. *Lodicules* are small but conspicuous.

This grass is common in water-logged situations, in paddy fields and in irrigated dry lands. Sometimes on the blades of this grass purple bands are present and the internodes and the spikes also become purplish.

It is really a weed of cultivation met with generally on rich soils. This grass is considered to be one of the best fodder grasses in India. All kinds of cattle eat it greedily.

Distribution. — It is found throughout India up to 6,000 feet and also in all warm countries.

[Pg 82]

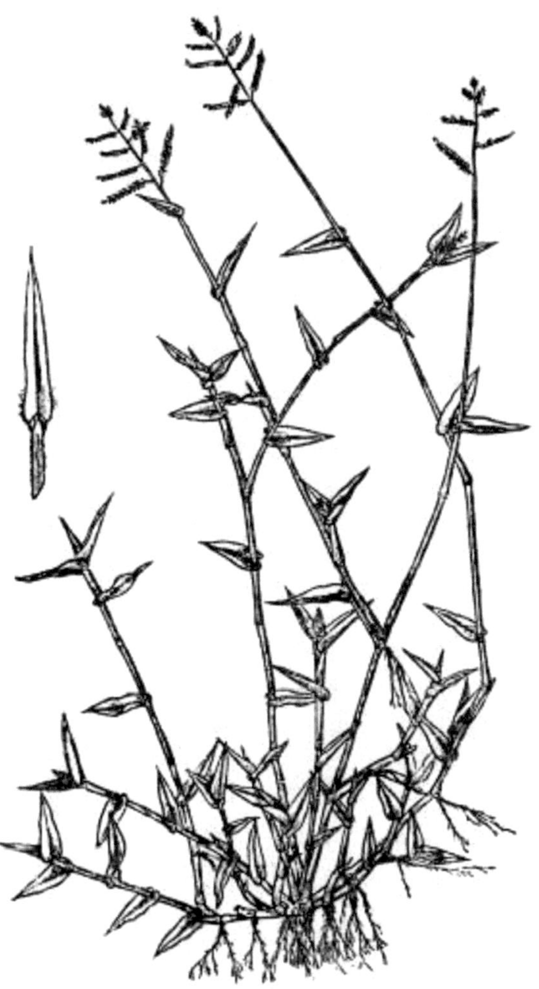

Fig. 91.—Panicum prostratum.

[Pg 83]

Panicum prostratum, *Lamk.*

The plant is a slender annual and it consists of several branches, prostrate and creeping, with adventitious roots at the nodes below, branching or ascending above, all green or sometimes purple above and green below, 4 to 18 inches long.

The *leaf-sheath* is striate, 1 to 2 inches long, glabrous or very sparsely hairy, purplish above and green below or all green, keeled, margins ciliate on one side only throughout its length. The *ligule* is a fringe of white hairs. The *nodes* are glabrous or pubescent.

The *leaf-blade* is short or long, varying from 1/2 to 2-1/2 inches in length and 3/16 to 5/16 inch in breadth, convolute when young, lanceolate to broadly ovate-lanceolate, acute or acuminate, upper surface glabrous, and the lower glabrous or with a few scattered tubercle-based hairs; margins are very minutely serrate; base is cordate, amplexicaul with a few long slender hairs (sometimes tubercle-based), just close to the white patch on both sides on the margin of the blade about the ligule. The midrib is distinct.

The *inflorescence* consists of five to fifteen or twenty spikes spreading in all directions, distant or crowded; peduncle varies from 1 to 4 inches. *Spikes* are 1/2 to 1-3/8 inches, sessile or shortly stalked; the *rachis* of the spike is slender, trigonous and scaberulous.

Fig. 92. — Panicum prostratum.
A. Front and back view of spike; B. front and back view of a spikelet; 1, 2, 3 and 4, the first, second, third and the fourth glume, respectively; 3a and 4a. the palea of the third and the fourth glumes; 5. anthers, ovary and lodicules.

The *spikelets* are crowded all on one side, 2- to 3-seriate, ellipsoidal, 1/20 to 1/16 inch long, glabrous or pubescent, pale green or purple on one side, in pairs on pedicels, one with a slightly longer pedicel than the other; fine long hairs, varying in number from one to eight and longer than the spikelets, are found on the pedicels at their tips in some plants and not in others.

There are four *glumes* in the spikelet. The *first glume* is very short about 1/4 of the third or less, semilunar, membranous, hyaline, subtruncate, obtuse or acute, generally nerveless, but rarely, obscurely 1- to 3-nerved. The *second glume* is membranous, ovate, acute, glabrous or pubescent and 7-nerved. The *third glume* is of about the same length as the second, 5-nerved, always paleate, with [Pg 84] or without stamens; *palea* is broad, margins infolded, 2-nerved, obtuse and hyaline; when stamens are present the *lodicules* are very conspicuous. The *fourth glume* is slightly shorter than the third, oblong or elliptic, apiculate, minutely rugulose, thinly coriaceous, with bisexual flower; *palea* is similar to the glume in texture and markings. *Stamens* are three with yellowish anthers. *Lodicules* are small and fleshy.

This plant occurs widely as a weed of cultivation in black cotton as well as other kinds of soil and shows variation in its leaves and spikelets. In some plants growing in somewhat dry places the leaves are shorter and broader, and those in favourable situations have longer narrower leaves. The spikelets are either perfectly glabrous or pubescent and long hairs may or may not be present on the pedicels. As regards colour the whole plant is green or the exposed portions of stems and spikelets are purplish. This grass is liked by cattle and is one of the most nutritious of Indian fodder grasses.

Distribution. — Throughout India and Ceylon in the plains. Common in the Tropics.

[Pg 85]

Fig. 93.—Panicum javanicum

[Pg 86]

Panicum javanicum, *Poir.*

This is an annual and it branches freely and the branches are decumbent and rooting at the nodes at the base, and erect to some

142

extent at the free end, 1 to 2 feet long; the internodes are glabrous, thinly striate, shallowly channelled on one side.

The *leaf-sheath* is somewhat compressed and loose, covered with scattered long hairs, some of them being tubercle-based; the margin is ciliate on one side only. The *nodes* are pubescent with long hairs. The *ligule* is a distinct fringe of hairs.

The *leaf-blade* is broadly lanceolate, cordate at base, amplexicaul, acuminate or acute, with scattered long hairs both above and below, and some of the hairs of the under surface are tubercle-based, convolute when young; margin of the leaf is wavy, minutely serrate, and ciliated with distant hairs towards the lower half of the leaf when young; the midrib is prominent below.

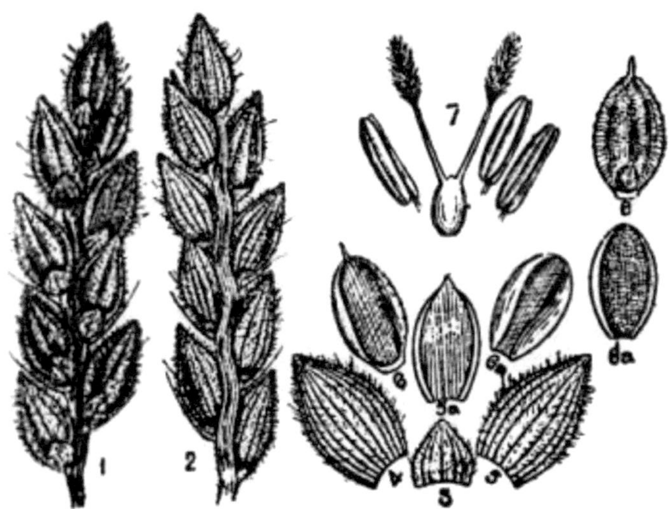

Fig. 94. — Panicum javanicum.
1 and 2. Front and back view of a spike; 3, 4, 5 and 6. the first, second, third and the fourth glume, respectively; 5a and 6a. the palea of third and fourth glumes; 7. the ovary and the stamens.

The *inflorescence* is a panicle of spikes on a short or long erect slender peduncle. *Spikes* vary from two to ten in number and in length from 1/2 to 2 inches, distant and spreading; the rachis of the spike is zigzag, somewhat flattened with a wavy ridge, scaberulous or glabrous, swollen towards the base and the swollen part is pubescent.

The *spikelets* are biseriate, loosely imbricate, ovate, acute, pubescent or villous (sometimes quite glabrous), sessile or shortly pedicelled; the pedicels have one or two (rarely more) long hairs.

There are four *glumes*. The *first glume* is small, membranous, less than 1/2 of the third glume, ovate, acute or obtuse, 3- to 5-nerved. The *second glume* is nearly equal to the third, ovate acute, generally 7-nerved and sometimes 7- to 13-nerved. The *third glume* is similar to the second in shape, generally 5-nerved and occasionally [Pg 87] 7-nerved, paleate with three stamens or empty; *palea* 2-nerved, ovate or oblong, margins infolded. The *fourth glume* is ovate or oblong, rugulose, chartaceous, apex with a distinct mucro concealed in the second and third glumes; *palea* same as the glume in texture, etc. *Anthers* are yellowish; *stigmas* are feathery and purple in colour; *lodicules* are small and fleshy.

This is an excellent fodder grass. Though it is an annual it grows rapidly under favourable conditions. A single plant found growing in the compound of the Agricultural College, Coimbatore, weighed 15 lb. and occupied 15 square feet of the ground. It flourishes in cultivated dry fields and in rich loamy soils. (See fig. 7.)

Distribution. — Plains of India and Ceylon and in Tropical countries generally.

[Pg 88]

Fig. 95.—Panicum ramosum.

[Pg 89]

Panicum ramosum, *L.*

This is an annual with stems erect or ascending from a creeping base, rooting at the lower nodes, 1 to 2 feet long. The stem is slender or stout, usually glabrous though occasionally glabrescent or pubescent, channelled on one side, branched from base upwards, and leafy.

The *leaf-sheath* is finely striate, keeled, thinly pubescent with the margins ciliate near the ligule. The *ligule* is only a fringe of short hairs. *Nodes* are softly hairy.

The *leaf-blade* is flat, linear-lanceolate, acuminate, softly pubescent or glabrescent on both the surfaces, with rounded or subcordate base and margins minutely serrate and ciliate, 2 to 6 inches long 1/6 to 1/2 inch broad; the midrib is distinct though slender with four to six main veins on each side.

The *inflorescence* is a pyramidal panicle 2 to 6 inches long, consisting of usually five to ten (rarely also up to twenty) erect or spreading spikes. *Spikes* are distant, alternate and in some the lower ones are opposite, 1/2 to 2-1/2 inches long or shorter. The *rachis* of the spike is thin, angular and scaberulous.

The *spikelets* are usually pubescent, ovoid or obovoid, acute, turgid, 1/8 inch, pale green and some occasionally purplish on one side, alternate close or distant, in pairs lower down and then one with a somewhat longer pedicel, solitary in the upper portions, pedicels with hairs, some of them especially those near the apex being longer.

Fig. 96.—Panicum ramosum.
1 and 2. Back and front view of spike; 3 and 4. front and back view of a spikelet; 5 and 6. first and second glumes; 7 and 8. third glume and its palea; 9 and 10. fourth glume and its palea; 11. ovary, anthers and lodicules.

There are four *glumes*. The *first glume* is nearly half the length of the third glume, broadly ovate, subacute, margin overlapping at the base, and usually 5-nerved. The *second glume* is broadly ovate acute, rather cuspidate, usually 5-nerved (rarely 7-nerved). The *third glume* is similar to the second glume, 5-nerved, paleate, empty; *palea* is hyaline oblong, acute. The *fourth glume* is ovoid-oblong, [Pg 90] acute, coriaceous, rugulose, with short broadened stipes, and three faint nerves; *palea* similar to the glume in texture and markings. *Anthers* are orange-yellow; *style* branches are purple. *Lodicules* are small and fleshy.

This grass is a common weed found in dry cultivated fields and open waste places and is one of the best fodder grasses available.

Distribution.—Plains throughout India and in Afghanistan.

[Pg 91]

Fig. 97.—Panicum distachyum.

[Pg 92]

Panicum distachyum, *L.*

This grass is an annual. Stems are slender, rarely stout, creeping and rooting at the nodes, pale green or purplish, with erect or ascending slender branches, varying in length from 10 to 15 inches, glabrous or pubescent, channelled near the nodes.

The *leaf-sheath* is glabrous or glabrescent and sometimes hirsute; margin is ciliate. The *ligule* is a fringe of short hairs. *Nodes* are glabrous or pubescent.

The *leaf-blade* is lanceolate or narrowly lanceolate, base cordate and subamplexicaul, glabrous or rarely sparsely hairy on both sides; margins are wavy here and there, finely serrate with tubercle-based hairs towards the base, the midrib is slender, not prominent and veins not distinct. There is considerable variation in leaves especially in the length. In the ordinary form it varies from 1/2 to 3 inches and even up to 6 or 7 inches sometimes in length and the breadth from 1/8 to 1/4 inch. In one form which is separated as a variety (var. *brevifolium*, Wight and Arnott,) the leaves are always short and broad, ovate-lanceolate never exceeding 1 inch in length.

The *inflorescence* consists of two or three, very rarely four erect or spreading distant spikes on a somewhat slender very hairy peduncle. *Spikes* are from 1/2 to 2 inches; *rachis* is slender, flexuous, flattened, scaberulous, with a few long hairs scattered singly along the margins or without these hairs.

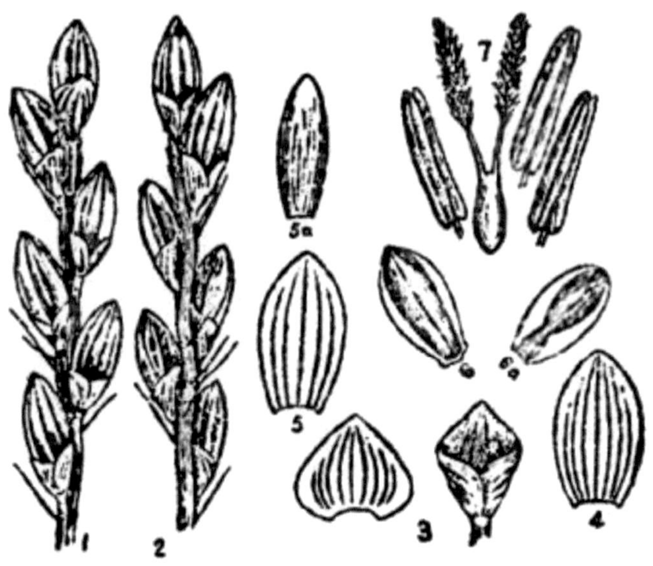

Fig. 98.—Panicum distachyum.
1 and 2. Front and back view of a portion of a spike; 3, 4, 5, and 6. the first, second, third and the fourth glume, respectively; 5a and 6a. palea of the third and the fourth glume, respectively; 7. anthers and ovary.

The *spikelets* are glabrous, ovate-oblong, acute, 1/8 inch, 1- or 2-seriate, subsessile, pale green, occasionally purplish on one side.

There are four *glumes*. The *first glume* is membranous, broadly ovate, obtuse with margins overlapping at the base, hardly half the length of the third glume, usually 5-nerved but occasionally 7-nerved. The attachment of the first glume is not close to that of [Pg 93] the second glume but is far lower. The *second glume* is ovate-acute, 7-nerved. The *third glume* is equal to the second, 5-nerved, paleate, empty; the *palea* is narrow, hyaline, acute. The *fourth glume* is ellipsoidal, obtuse, chartaceous, minutely and obscurely rugulose, faintly 3-nerved, with the base somewhat thickened. *Palea* is similar to the glume in texture. *Anthers* are orange-yellow. *Lodicules* are minute and fleshy. Style branches are purple.

150

This grass is fairly common in open and loamy and sandy soils. The form (var. *brevifolium*, Wight & Arnott) is fairly common in Coimbatore District.

Distribution. — Plains of India and Ceylon. Not recorded from the Bombay Presidency. It occurs in China, Malaya and Australia.

[Pg 94]

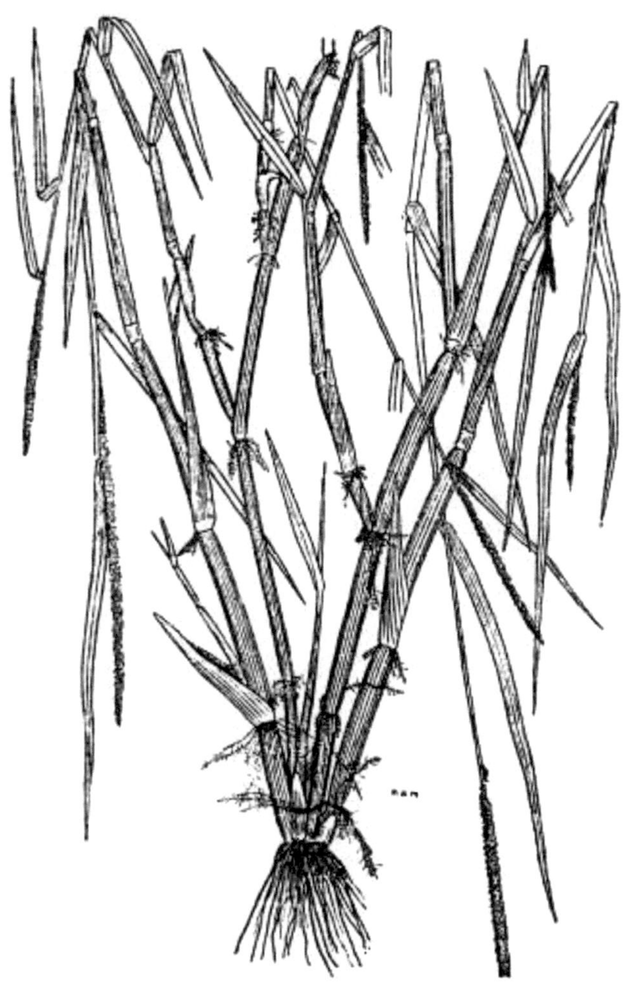

Fig. 99.—Panicum interruptum.

[Pg 95]

Panicum interruptum, *Willd.*

This is a large perennial grass with stems reaching 5 to 6 feet in length, flourishing in marshes and in the edges of ponds and tanks.

The stems are long, stout and spongy below, ascending from a creeping and rooting or floating root-stock; the lower internodes are often 1/2 inch or more in thickness, with nodes bearing in fascicles long stout roots clothed with fine lateral roots; and the upper internodes are long and slender.

The *leaf-sheath* is glabrous, striate. The *ligule* is a short broad membrane.

The *leaf-blade* is soft, flat, many-nerved, linear, finely acuminate, margins smooth, base rounded or subcordate, glabrous, 6 to 12 inches long, 1/4 to 1/2 inch broad.

The *inflorescence* is a strict spike-like panicle, 6 to 12 inches long by 1/4 to 1/3 inch broad, cylindric, interrupted below; the rachis terete, stout, channelled.

The *spikelets* are glabrous, green, herbaceous, densely packed in small fascicles, ovoid lanceolate, 1/6 to 1/5 inch long; many spikelets are imperfect.

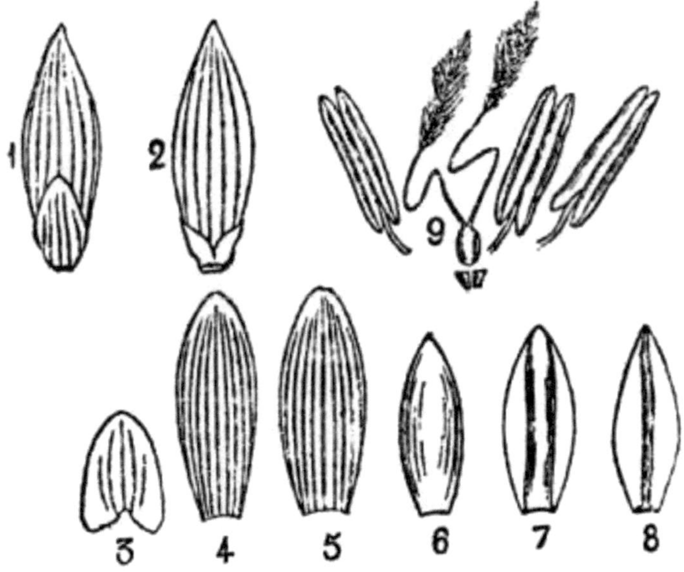

Fig. 100.—Panicum interruptum.
1 and 2. Front and back view of a spikelet; 3. first glume; 4. second glume; 5. third glume; 6. palea of third glume; 7 and 8. the fourth glume and its palea; 9. ovary, lodicules and stamens.

There are four *glumes*. The *first glume* is hyaline, membranous, about 1/3 the length of the third glume, broadly ovate or orbicular, obtuse, 5-nerved. The *second glume* is membranous, ovate-oblong, obtuse, prominently 9-nerved. The *third glume* is as long as the second but broader, ovate-oblong, 9-nerved, paleate; *palea* is small with three stamens or without them. The *fourth glume* is shorter than the third glume, lanceolate, subacute, thinly coriaceous white, polished, dorsally convex; the *palea* is as long as the glume and thinly coriaceous. There are two small *lodicules*.

This is a rank marsh grass growing abundantly in permanent marshes and edges of tanks and ponds. Cattle eat this along with other grasses, when young and not covered with algæ.

Distribution.—In swampy situations throughout India and Ceylon.

[Pg 96]

Panicum trypheron, *Schult.*

The plant is a tufted annual leafy at the base, with branches spreading a little at the base and then erect, varying in length from 1/2 to 3 feet. Stems are stout or slender, cylindric or slightly compressed towards the base.

The *leaf-sheath* is striated, green or purple tinged, shorter than the internodes, the upper portion hairy (sometimes tubercle-based) and the lower glabrous, with sometimes ciliate margin. The *ligule* is a short membrane with a fringe of slender hair-like processes. *Nodes* are glabrous.

The *leaf-blade* is flat, convolute when young, linear-lanceolate, acute or narrow linear-lanceolate, acuminate, hairy on both sides (hairs indistinctly bulbous-based); margin is very minutely serrate and often ciliate with tubercle-based hairs; base is narrowed, slightly rounded or cordate; midrib is conspicuous though narrow and keeled; length 1 to 7 inches and breadth 1/8 to 3/8 inch.

The *inflorescence* is a diffuse panicle 4 to 14 inches long with filiform, divaricate, scaberulous, angled branches; the main *rachis* is angular, smooth below and scaberulous above; peduncle is cylindric, striate, 2 to 12 inches long. Branches are irregularly distantly alternate, solitary or rarely two, swollen at base, dividing into slender filiform spreading branchlets; the lower branches from 3 to 7 inches in length and getting shorter upwards. Branchlets are 1/2 to 3 inches, capillary, angular and further dividing.

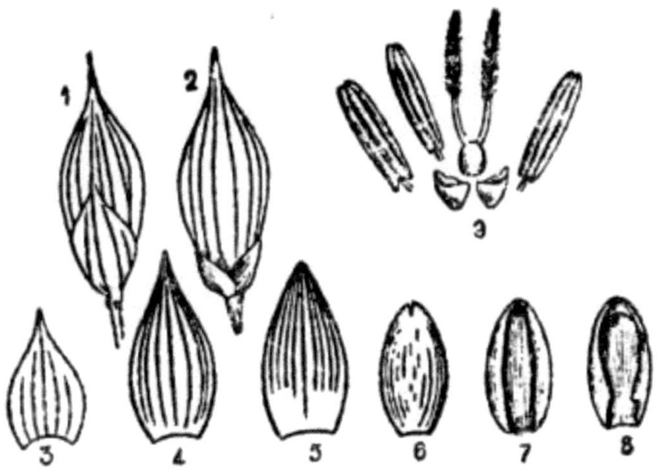

Fig. 101.—Panicum trypheron.
1 and 2. Front and back view of the spikelet; 3, 4 and 5. the first,
second and the third glume, respectively; 6. palea of the third
glume; 7 and 8. the fourth glume and its palea; 9. stamens, ovary
and lodicules.

The *spikelets* are ovate, acuminate, binate (sometimes solitary or
three) on a common finely filiform stalk, one long and the other
short pedicelled, pale or yellowish green, or purple; pedicels are
angular, scabrid or scaberulous, slightly swollen at the top and
sometimes with setose hairs also.

There are four *glumes*. The *first glume* is green or purple, broadly
ovate, acuminate, clasping at the base, about two-thirds of the third
glume, membranous, nerves five, the lateral two stout and [Pg 97]
anastomosing halfway, finely scaberulous especially on the nerves
and more so on the central one. The *second glume* is slightly longer
than the third, green or purple, ovate, acuminate, generally 7-
nerved and sometimes also with two more indistinct marginal
nerves, i.e., 9-nerved, scaberulous on the nerves. The *third glume* is
pale green or yellow, ovate-oblong, acute or subacute, obscurely
scaberulous, 9-nerved (two of the nerves in the middle sometimes
not running to the base), paleate, empty. *Palea* is hyaline, smaller
than the glume, oblong, obtuse, minutely two-lobed or two-toothed

at the apex; margins broadly infolded. The *fourth glume* is elliptic obtuse, shorter than the third, smooth, shining, coriaceous, dorsally convex, with a prominent short, broad stipe at the base which is persistent with the glume, 5-nerved, sometimes with seven nerves especially when young (two marginal ones being indistinct). *Palea* is similar to the glume in texture. *Anthers* are three, linear, orange yellow. *Lodicules* are two and prominent though small. *Stigmas* feathery and white.

P. tenellum, Roxb. Fl. Indica I. 306 is probably not this plant though quoted as a synonym, for it is described as having culms prostrate and rooting at the nodes.

This grass is of wide distribution in the Presidency, but it is nowhere abundant. It is fairly common in cultivated dry fields. Cattle like this grass.

[Pg 98]

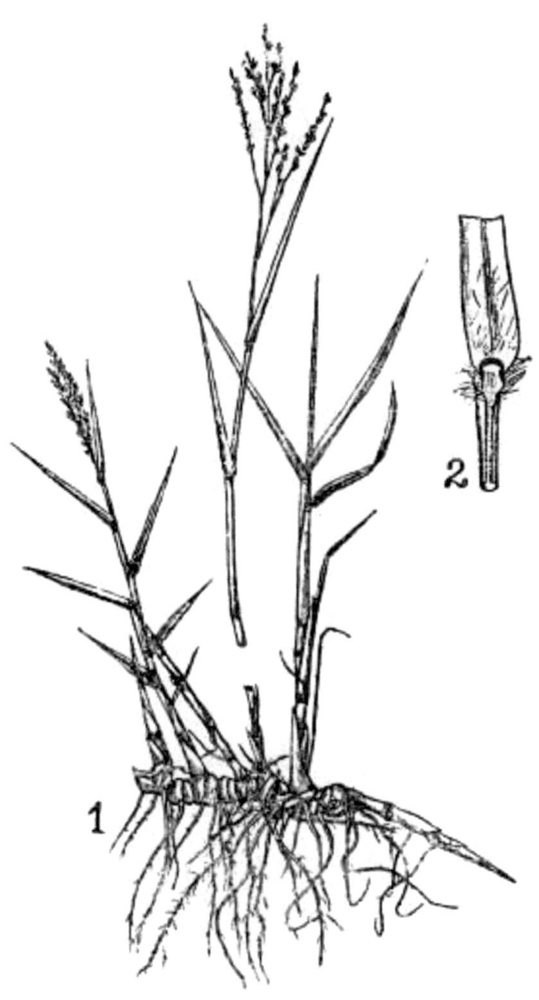

Fig. 102.—Panicum repens.
1. Full plant; 2. a portion of the leaf and ligule.

[Pg 99]

Panicum repens, *L.*

This is a perennial glaucous grass with stoloniferous and rhizomiferous stems bearing ordinary erect leafy branches, and the branches come out piercing through the leaf-sheath (extravaginal).

Stems are numerous, stiff and erect, 1/2 to 3 feet in length, glabrous, covered below by brownish or whitish scale-leaves, and above with densely distichous leaves.

The *leaf-sheath* is firm, distinctly striate, glabrous, margins ciliate on both sides up to the point of overlapping and then the outer margin alone ciliate. The *ligule* is a short thin membrane with very short cilia on the free margin. The *nodes* are glabrous.

The *leaf-blade* is glaucous, narrow, lanceolate, thinly coriaceous, acuminate with a hardened tip, 1 to 7 or 9 inches long, 1/2 to 1/4 inch broad, flat or involute when slightly faded, with a few distantly scattered hairs above, especially towards the lower portion of the blade when young, and becoming glabrous later, glabrous on the lower surface, margin is finely serrate and with a few cilia towards the base, some hairs being tubercle-based; base of the blade is rounded or cordate, midrib is prominent and keeled.

Fig. 103. — Panicum repens.
1. Spike; 2 and 3. front and back view of a spikelet; 4, 5 and 6. first,

second and third glumes; 7. palea of the third glume; 8 and 9. fourth glume and its palea; 10. lodicules, stamens and ovary; 11. leaf showing ligule.

The *inflorescence* is a panicle, contracted and not much exserted from the topmost leaf-sheath, 3 to 8 inches long, branches are usually many, erect, the lower being 2 to 5 inches long, slender, angular and scaberulous.

The *spikelets* are glabrous, erect, pale or pale green, sometimes purplish also on one side, ovate-oblong or oblong-lanceolate, acute, 1/8 inch, pedicels are long with cupular tips.

There are four *glumes* in the spikelet. The *first glume* is hyaline, broadly ovate, rounded and shortly acute or subacute, indistinctly 3- to 5-nerved or nerveless, less than one-third of the height of the third glume. The *second glume* is membranous, ovate-lanceolate [Pg 100] acute, 7- to 9-nerved. The *third glume* is equal to and broader than the second, always paleate and with three stamens and 9-nerved; *palea* is hyaline, oblong, obtuse or subacute, margins folded. The fourth *glume* is white, coriaceous, smooth and shining, oblong, acute, shortly and broadly stipitate, with the margins folded inwards exposing only a third of the palea; *palea* is similar to the glume in texture and marking. *Anthers* are deep orange in colour. *Lodicules* are distinct though small. *Stigmas* are deep purple when mature, and pale when young.

This grass flourishes in moist situations such as the bunds of paddy fields, tank beds and edges of marshes and is an excellent binder of the soil. When once established it is very difficult to get rid of it, on account of its rhizomes. Owing to the resemblance of the rhizomes to ginger, some call this grass Ginger-rooted grass. Cattle are fond of this grass.

Distribution.—Throughout India, but not so common on the West and not recorded from Bombay. It is said to occur in South Europe, Australia, North Africa and Brazil.

[Pg 101]

5. Chamæraphis, *Br.*

These are glabrous marsh or aquatic grasses. Leaves are linear or lanceolate. The inflorescence is a panicle. The spikelets are one-to two-flowered, subsessile and subsecund on the branches which are produced as awn-like bristles beyond the ultimate spikelet, obscurely jointed and persistent on their obconic short pedicels, narrowly lanceolate and terete. The spikelet consists of four glumes. The first glume is very small, hyaline, suborbicular, nerveless and truncate. The second glume is the longest, green, membranous, narrowly lanceolate, acuminate or narrowed into a rigid awn, 7- to 11-nerved. The third glume is lanceolate, acute, or aristately acuminate, 7-nerved, paleate, male or neuter, the palea is smaller than the glume and hyaline. The fourth glume is much smaller than the third, stipitate, bisexual or female, oblong or ovate-oblong, acute, flat, thinly coriaceous, nerveless and paleate; the palea is hyaline, as broad as the glume, acute and nerveless. The lodicules are cuneate. Stamens are three. Stigmas are laterally exserted. Grain is oblong, compressed.

[Pg 102]

Chamæraphis spinescens, *Poir.*

A glabrous aquatic or marsh grass, with much branched floating stems. Stems are leafy, elongate, ascending, varying in length from 1 to 3 feet.

The *leaf-sheaths* are long, smooth, loose, with naked margins. The *ligule* is a ridge of hairs. The *nodes* are glabrous.

The *leaf-blade* is flat, narrowly linear-lanceolate, smooth or scabrid, acuminate, base narrowed, 1 to 3-1/2 inches long and 1/16 to 1/8 inch wide.

The *inflorescence* is a pyramidal panicle, contracted or diffuse, with a leaf very near its base; peduncle is short; branches of the panicle, filiform, angular, flexuous, bearing one or more spikelets and produced as a bristle beyond the last spikelet.

The *spikelets* are 1/6 to 1/4 rarely 1/3 inch long including the awn, subsessile and somewhat on one side on the branches, obscurely articulate but persistent on the pedicels, pale or green, lanceolate.

There are four glumes in the spikelet. The *first glume* is hyaline, suborbicular, rounded at the tip and nerveless, 1/30 inch or less. The *second glume* is membranous, lanceolate, smooth or setosely scabrid on the sides, 9- to 11-nerved, with a long scabrid awn which is sometimes as long as the body of the glume. The *third glume* is shorter than the second, finely acuminate, or awned, 7-nerved, membranous, paleate and with three *stamens* and two *lodicules*; the *palea* is shorter than the glume, linear-oblong, subacute. The *fourth glume* is ovate-lanceolate, nerveless, acute, paleate with three *stamens*, *ovary* and two *lodicules*; *palea* is hyaline, narrow, quarter the length of the third glume. Grain is obovate oblong.

Fig. 104.—Chamæraphis spinescens.
1. Terminal portion of a spike showing the bristle; 2, 3, 4 and 6. the

first, second, third and the fourth glume, respectively; 5. palea of third glume with its anthers and lodicules; 7. palea of the fourth glume; 8. ovary; 9. lodicules.

Distribution.—This plant is found at the edges in ponds, tanks and marshes all over the Presidency.

[Pg 103]

6. Spinifex, *L.*

This is a stout, rigid, much branched, gregarious and dioecious grass, flourishing in sand on the sea coast. Leaves are long, narrow rigid, involute, spreading and recurved and thickly coriaceous. Male spikelets are 1- to 2-flowered, subsessile, distichous, jointed on rigid peduncled spikes, which are collected in umbels and surrounded by spathaceous leafy bracts. The spikelets have four glumes. The first two glumes are empty. The third and the fourth paleate and triandrous and sometimes the former is empty. Female spikelets are collected in large globose heads of stellately spreading very long rigid rod-like processes surrounded by shorter subulate bracts. Each spikelet is solitary, and articulate at the very base of a rachis, lanceolate, 1-flowered. There are four glumes. The first three glumes are as in the male spikelets, but larger. The third is paleate, empty. The fourth glume has a female flower. The lodicules are large and nerved. Styles are long, free, with short, feathery stigmas. Grain free within the hardened glumes.

[Pg 104]

Spinifex squarrosus, *L.*

A perennial littoral dioecious grass forming bushes. Stems are glaucous, smooth, solid, woody, thick below, freely branching, 5 to 10 feet long or more.

The *leaf-sheath* is smooth, imbricating, 1/2 to 1-1/2 inches long. The *ligule* is a row of stiff long hairs.

The *leaf-blade* is narrow, rigid, thickly coriaceous, concavo-convex tapering from the base to the tip, spreading and recurved, 4 to 6 inches long.

The *male inflorescence* consists of several spikes, 1 to 3 inches long, forming umbels, with membranous leafy spathaceous bracts which are shorter than the spikes.

The *spikelets* are usually 2-flowered, smooth, articulate on short peduncles, distichous, 1/3 to 1/2 inch long.

There are four *glumes*. The *first glume* is shorter than the second, ovate, obtuse, 7- to 9-nerved. The *second glume* is similar to the first, but longer. The *third* and the *fourth glumes* are longer than the second glume, 5- to 7-nerved, paleate and triandrous; *palea* of both are lanceolate with ciliate keels.

Fig. 105.—Spinifex squarrosus.
Male plant—1. A branch with the male inflorescence; 2. a spike; 3. a spikelet; 4, 5, 6 and 7. the first, second, third and the fourth glume, respectively; 6a. palea of the third glume; 6b. extra palea like structure found occasionally in the palea of the third glume; 7a. palea and lodicules of the fourth glume.

The *female inflorescence* is a large globose head consisting of short spikelets articulate at the very base of the rachis, short bracts and very long, spreading, rigid rod-like rachises. The *spikelets* are solitary with four glumes and 2-flowered. The *first glume* is oblong-lanceolate, many-nerved, longer than the other glumes. The *second glume* is shorter, 7-nerved. The *third glume* is empty, 5-nerved. The *fourth glume* is ovate-lanceolate and abruptly narrowed above the middle, 5-nerved and paleate, palea is shorter than the glume but broader, 2-nerved and acute. *Lodicules* are two, large, cuneate at base and strongly nerved. *Stigmas* are oblong. Grain is clavate and tipped by the style base.

[Pg 105]

This grass grows luxuriantly in the sands near the sea on both the coasts of the Madras Presidency.

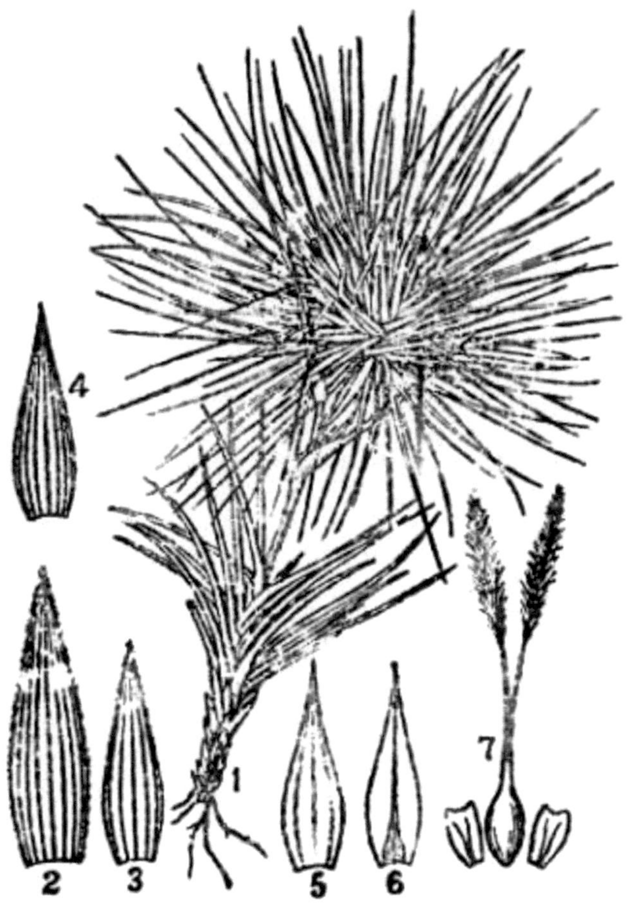

Fig. 106.—Spinifex squarrosus.
Female plant—1. A branch with female inflorescence; 2, 3, 4 and 5. the first, second, third and the fourth glume, respectively; 6. palea of the fourth glume; 7. ovary and the lodicules.

Distribution.—Throughout the sandy coasts of India and Ceylon.

[Pg 106]

7. Axonopus, *Beauv.*

These are annual or perennial grasses. Inflorescence is a panicle consisting of digitate or whorled, slender or stout spike-like racemes. Spikelets are solitary, binate or fasciculate, 2-flowered, jointed on the pedicel and awned. There are four glumes. The first glume is the shortest, ovate, acuminate, aristate or cuspidate, hyaline, glabrous and 3-nerved. The second glume is ovate or ovate-lanceolate, acuminate or awned, 5-nerved, lateral nerves being marginal and hairy. The third glume is oblong or oblong ovate, acute, 5-nerved, paleate, male; palea is very short and small, bipartite. The fourth glume is as long as the third and the second, oblong or ovate, coriaceous, narrowed into a straight terminal awn, paleate and bisexual; palea is oblong, coriaceous and 2-nerved. Lodicules are cuneate. Stamens are three with linear anthers. Stigmas are linear, laterally exserted. Grain is oblong, free within the hardened glume and its palea.

[Pg 107]

Fig. 107.—Axonopus cimicinus.

[Pg 108]

Axonopus cimicinus, *Beauv.*

It is a perennial grass. Stems are tufted, erect or slightly decumbent at the base, 1 to 2 feet long.

The *leaf-sheath* is distinctly striate, covered with scattered long tubercle-based hairs, very rarely glabrous, keeled. The *ligule* consists
of a row of hairs. The *nodes* are hairy.

The *leaf-blade* is flat, ovate-lanceolate, broad and cordate at base,
subacute or obtuse, with a distinct midrib and three main veins on
each side of it, glabrous on both sides, but usually with tubercle-
based hairs on the two sides of the midrib, on the lower side, the
margins are distinctly ciliate with tubercle-based long stiff hairs and
very finely serrate; the blade varies in length from 3/4 to 3 inches
and in breadth from 3/4 to 1/2 inch.

The *inflorescence* consists of three to ten spikes springing from the
top of a slender glabrous peduncle 2 to 6 inches long. The *spikes* are
whorled, about 3 inches or so in length, naked towards the base to
about one-fourth of its length, the rachis is fine, filiform, scabrid.

The *spikelets* are solitary or binate, dorsally compressed, pale
green or reddish, very shortly pedicelled, 1/4 to 5/16 inch long
inclusive of the short awn, pedicel is cupular at the tip.

There are four *glumes* in the spikelet. The *first glume* is somewhat
narrow ovate-lanceolate, hyaline, acuminate and 3-nerved. The
second glume is membranous, ovate-lanceolate, twice as long as the
first glume, cuspidately acuminate, 5-nerved; the two marginal
nerves are provided with long reddish bristly hairs. The *third glume*
is oblong lanceolate, obtuse, 5-nerved, a little shorter than the second glume, paleate and with stamens; *palea* is short. The *fourth
glume* is coriaceous, ovate-lanceolate, nearly as long as the second
glume, awned at the apex, paleate, with three stamens and an ovary; the *palea* is as long as the glume, elliptic oblong, obtuse. *Lodicules*
are small, cuneate.

Fig. 108. — Axonopus cimicinus.
1. A portion of the spike showing spikelets; 2, 3, 4 and 5. the first, second, third and the fourth glume, respectively; 4a and 5a. the palea of the third and the fourth glume, respectively; 6. lodicules, stamens and the ovary.

This is a common grass growing in the plains and lower hills in waste places.

Distribution. — Occurs all over India.

[Pg 109]

8. Setaria, *Beauv.*

These are usually annuals. Inflorescence is usually a spike-like panicle. Spikelets are 1- to 2-flowered, jointed on very short pedicels which bear persistent scabrid or barbed bristles (modified branchlets). There are four glumes. The first glume is the shortest, equal to about half the length of the third, membranous, 3- to 5-nerved. The second glume is equal to or shorter than the fourth, 5- to 7-nerved. The third glume more or less exceeding and resembling the second glume, neuter, rarely paleate and male. The fourth glume is coriaceous or crustaceous, plano-convex, bisexual, 5-nerved and paleate;

palea is as long as the glume. Lodicules are broadly cuneate. Stamens are three. Stigmas are laterally exserted. Grain is tightly enclosed by the hardened glume and its palea and is oblong or ellipsoid.

KEY TO THE SPECIES.

- Bristles with spreading or erect barbs.
    - o Inflorescence cylindric, continuous and not interrupted, with six to twelve bristles in the involucel 1. S. glauca.
    - o Inflorescence interrupted, with three to six bristles in the involucel 2. S. intermedia.
- Bristles with reversed barbs 3. S. verticillata.

[Pg 110]

Setaria glauca, *Beauv.*

This is a tufted annual grass. Stems are slender, simple or branched, erect or ascending.

The *leaf-sheaths* are glabrous. *Nodes* are glabrous and sometimes the lower are rooting. The *ligule* is a fringe of long hairs.

The *leaf-blades* are lanceolate-linear, flat, finely acuminate, with a rounded base and very finely and minutely serrate margin, glabrous on both the surfaces or occasionally sparsely hairy on the upper surface and varying in length from 4 to 12 inches or more, and in breadth from 1/4 to 1/3 inch.

The *inflorescence* is a cylindric, densely flowered, spike-like raceme, 1 to 4 inches long, usually yellow, rarely purplish or pale green, the bristles of involucels vary from six to twelve and are pale or reddish brown, 1/6 to 1/3 inch long with fine erect or spreading barbs.

The *spikelets* are numerous and are very closely set along the rachis of the inflorescence, 1/8 inch long, glabrous and ellipsoidal.

There are four *glumes* in the spikelet. The *first glume* is less than half the length of the third glume, broadly ovate, hyaline, 3-nerved.

The *second glume* is a little longer than the first but shorter than the third, broadly ovate or suborbicular, hyaline, 5-nerved. The *third glume* is longer than the second, as long as the fourth, membranous and 5-nerved, paleate, empty or with stamens. The *fourth glume* is coriaceous, broadly elliptic, obtuse, dorsally convex, transversely rugose, pale. The *anthers* are orange and the *styles* purple.

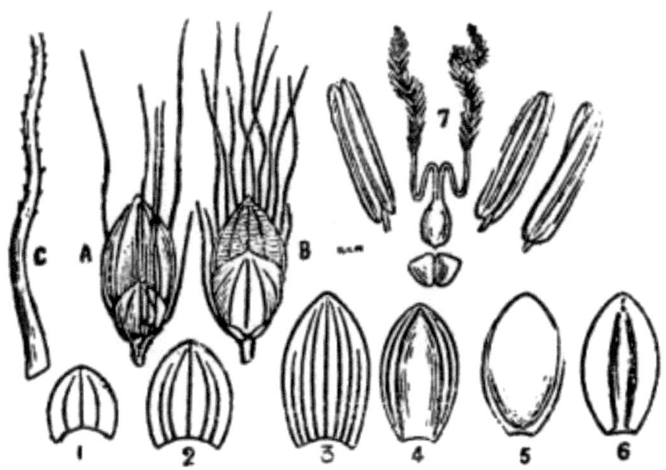

Fig. 109. —Setaria glauca.
A and B. spikelets; C. a bristle; 1, 2 and 3. The first, second and the third glume, respectively; 4. palea of the third glume; 5. the fourth glume; 6. palea of the fourth glume; 7. ovary, anthers and lodicules.

This is a fairly common grass especially in cultivated ground all over the Presidency, but not very widely distributed. Cattle are fond of this grass.

Distribution. — Throughout India.

[Pg 111]

Setaria intermedia, *R. & S.*

This is an annual with straggling, slender, erect or ascending stems, 2 to 3 feet long.

The *leaf-sheath* is glabrous, keeled, with the margins ciliate with long hairs. The *ligule* is a fringe of close set long hairs. The *nodes* are glabrous and the lower rooting.

The *leaf-blade* is linear-lanceolate, narrowed towards the base, finely acuminate, with fine hairs scattered on both the surfaces and with numerous long hairs at the mouth and with very finely serrate margins, varying in length from 2 to 8 inches or more, 1/8 to 3/4 inch in breadth.

The *inflorescence* is a narrowly pyramidal spike-like panicle, 4 to 6 inches long, the main rachis is glabrous and grooved, branches are short, crowded above, scattered and distant below, with close and densely set spikelets; the bristles of involucels are 1/4 inch long, slender, flexuous with erect barbs varying in number from three to six.

The *spikelets* are ovoid.

There are four *glumes* in the spikelet. The *first glume* is orbicular, oblong or ovate, about one-third the length of the third glume, hyaline, 3-nerved. The *second glume* is half as long as the third, broadly ovate, hyaline, 5-nerved. The *third glume* is as long as the fourth, broadly ovate, thinly membranous, 5-nerved, paleate, empty. The *fourth glume* is broadly ovate, or suborbicular, very concave, coriaceous, transversely rugulose, yellowish brown. *Anthers* are orange or yellow and *styles* purplish. *Lodicules* are very small.

Fig. 110.—Setaria intermedia.
1. A branch with spikelets; 2 and 2a. spikelets; 3, 4 and 5. the first, second and the third glume, respectively; 5a. the palea of the third glume; 6. the fourth glume; 6a. the fourth glume and its palea; 6b. palea of the fourth glume; 7. ovary, anthers and lodicules.

Fairly common in rich soils in sheltered places. Cattle are very fond of this grass as the leaves are flaccid and tender.

Distribution.—Probably all over India.

[Pg 112]

Fig. 111.—Setaria verticillata.

[Pg 113]

Setaria verticillata, *Beauv.*

This is an annual grass, with erect, ascending, stout or slender, leafy stems, more or less branched and varying in length from 1 to 5 feet.

The *leaf-sheaths* are smooth, glabrous. The *ligule* is a fringe of hairs. *Nodes* are glabrous.

The *leaf-blades* are thin, flat, glabrous, sparsely hairy and scaberulous, linear or linear-lanceolate, tapering to a fine point, base usually narrowed, 4 to 10 inches long and 1/4 to 3/4 inch broad.

The *inflorescence* is a spike-like or subpyramidal panicle, cylindric or oblong, coarsely bristly, 2 to 7 inches long, bristles one or few, studded with conspicuously reversed barbs or teeth, 1/6 to 1/3 inch long.

The *spikelets* are ellipsoidal, obtuse, glabrous, 1/12 inch long.

174

There are four *glumes*. The *first glume* is very small, broadly ovate, acute, hyaline, faintly 3-nerved. The *second glume* is as long as the spikelet or a little shorter, ovate, subacute, thinly membranous and 5-nerved. The *third glume* is equal to the second or a little longer, membranous and 5-nerved, paleate or empty, palea when present, is small and hyaline. The *fourth glume* is elliptic-oblong, plano-convex, subobtuse, smooth or shining, though faintly striate, coriaceous with incurved margins; *palea* is coriaceous, as long as the glume, elliptic, faintly striate. *Stamens* are three. *Lodicules* are small.

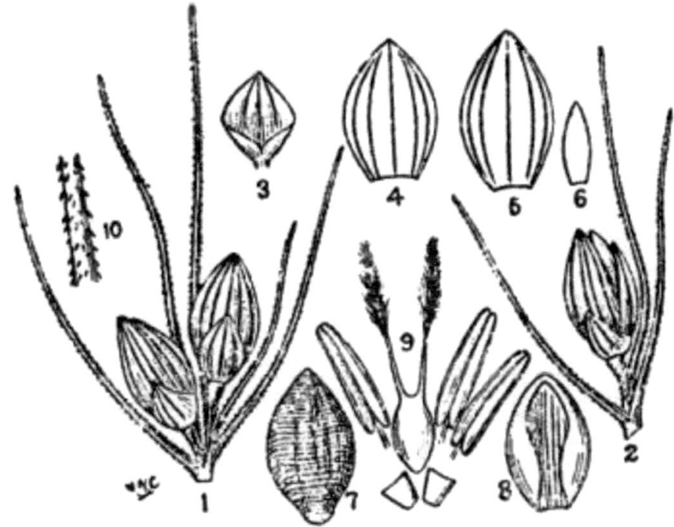

Fig. 112.—Setaria verticillata.
1 and 2. Spikelets with bristles; 3, 4 and 5. the first, second and the third glume, respectively; 6. palea of the third glume; 7 and 8. the fourth glume and its palea; 9. ovary, stamens and lodicules; 10. a bit of the bristle showing the reversed barbs.

This grass grows in shady places in very rich soils generally and is abundant in shady nooks and corners where there are rubbish heaps.

Distribution.—Throughout India and Ceylon.

[Pg 114]

9. Pennisetum, *Pers.*

These are annual or perennial grasses. Leaves are usually narrow. The inflorescence is a spike-like raceme consisting of involucellate clusters of shortly pedicellate spikelets, involucels consist of unequal, simple or branched bristles. Spikelets are obovoid or lanceolate, 1- to 2-flowered, persistent on their stalks, one to three in an involucel. There are usually four glumes in a spikelet. The first glume is minute or absent. The second glume is shorter than the third, membranous, 3- to 5-nerved, rarely wanting. The third glume is as long as the fourth, lanceolate, paleate or not, male or empty. The fourth glume is coriaceous, lanceolate, bisexual or female. There are three stamens with linear anthers. Styles long. Lodicules are small if present. Grain is oblong, free within the hardened fourth glume and its palea.

KEY TO THE SPECIES

- Bristles of the involucel slender and not dilated at the base, and free; leaves very long. 1. P. Alopecuros.
- Bristles of the involucel dilated below and connate at base. 2. P. cenchroides.

[Pg 115]

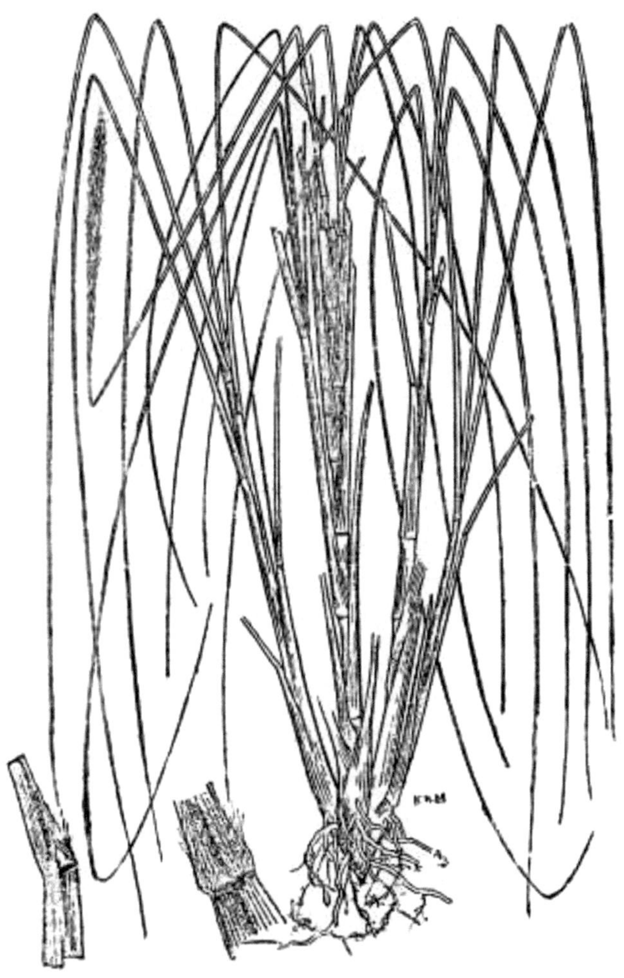

Fig. 113. — Pennisetum Alopecuros

[Pg 116]

Pennisetum Alopecuros, *Steud.*

This is a perennial grass, densely tufted and growing to a height of 2 to 3-1/2 feet. Stems are stout, erect and much branched above.

The *leaf-sheaths* are distichous, compressed, glabrous or rarely hairy.

The *leaf-blades* are convolute, narrow, linear, coriaceous, strongly keeled, glabrous but with tufts of soft hairs at the base, 12 to 18 inches long, 1/10 to 1/6 inch broad. The *ligule* is a ring of hairs.

The *inflorescence* is a spike-like raceme, varying in length from 5 to 7 inches. The involucels are shortly stalked, with a few unequal bristles which are free down to the base and two to three times as long as the spikelet.

Spikelets are lanceolate, acute, solitary, 3/8 inch long.

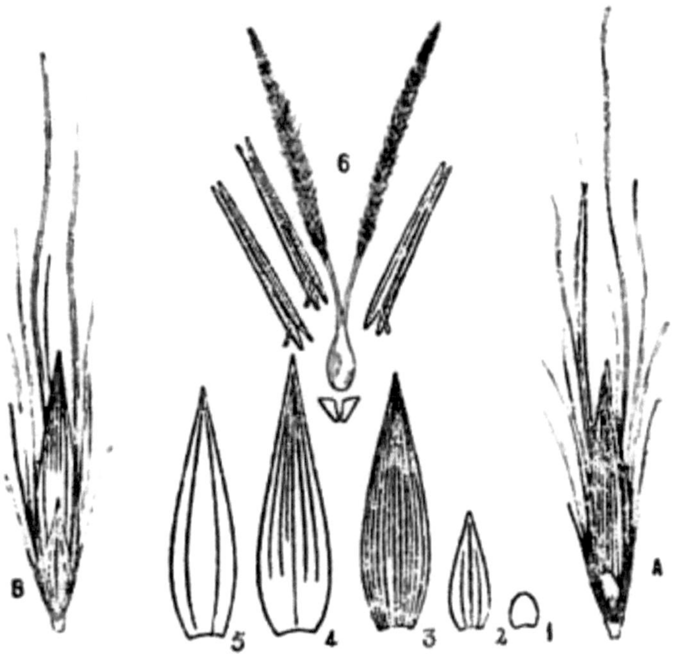

Fig. 114.—Pennisetum Alopecuros.
A and B. Spikelets front and back view; 1, 2, 3 and 4. the first, second, third and the fourth glume, respectively; 5. palea of the fourth glume; 6. the ovary, stamens and lodicules.

178

The *first glume* is very small, almost orbicular, hyaline and nerveless. The *second glume* is about 1/3 the length of the third glume, lanceolate, acuminate, 3-nerved. The *third glume* is about 1/3 inch long, lanceolate, acuminate, 7- to 11-nerved, epaleate and with infolded margins. The *fourth glume* is a little longer than the third, lanceolate, acuminate, with infolded margins 5- or 6-nerved, paleate and enclosing a complete flower. The *palea* is lanceolate, acuminate, as long as the glume. There are three *stamens* with long, narrow, yellow anthers. *Stigmas* are feathery. *Lodicules* are either absent or very minute.

This is a very coarse grass usually growing in stiff soils especially near wet places.

Distribution.—Occurs all over Southern India both on the plains and on low hills.

[Pg 117]

Fig. 115.—Pennisetum cenchroides.

[Pg 118]

Pennisetum cenchroides, *Rich.*

This grass is a perennial. It consists of aerial branches and underground rhizomiferous stems, bearing thick fibrous roots and numerous buds covered by scarious sheaths. The aerial branches are tufted, erect or decumbent and geniculately ascending when in flower, much branched from the base, 6 to 24 inches long (under favourable conditions may reach even 3 to 4 feet in length).

The *leaf-sheath* is slightly compressed, keeled, with scattered long hairs outside, shorter than the internodes. The *ligule* is a short thin membrane fringed with hairs.

The *leaf-blade* is linear, tapering to a very fine point 1-1/2 to 6 inches (sometimes 18 to 20 inches) by 1/8 to 1/4 inch scaberulous with fine long tubercle-based deciduous hairs scattered above, and the lower surface glabrous or with a few distantly scattered fine long hairs, broad at the base and constricted at the point of junction with the sheath.

The *inflorescence* is a raceme of spikes, varying from 1-1/2 to 3-1/2 inches, with the spikes mostly densely arranged, though occasionally distant and not close-set, on a long; slender, puberulous or scaberulous peduncle; *rachis* is flexuous, flattened, grooved and scaberulous. The *spikes* have involucels, consisting of two series of bristles, the outer bristles are horizontal or reflexed, numerous, fine, filiform, scabrid and purple above, shorter or longer than the spikelets; the inner bristles are two to three times longer than the spikelets, flattened and thickened at the base with a strong green nerve, ciliated with long tubercle-based hairs; one of the bristles is longer than the others and the bases of the bristles are connate at the very base into a ring; the upper portion of the bristles are filiform, scabrid and purple, the lower flattened portion being pale.

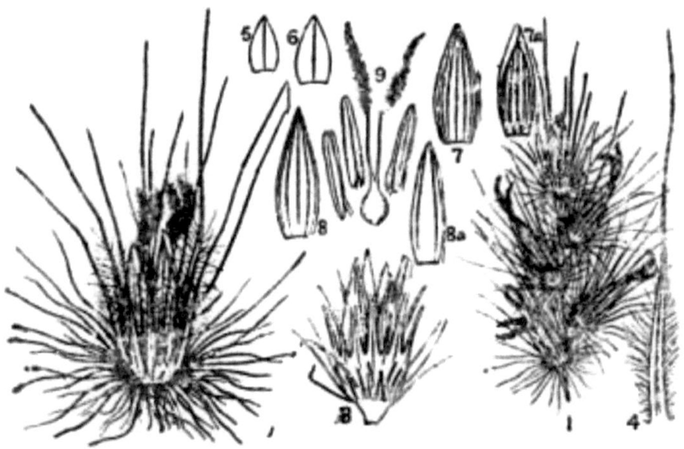

Fig. 116. — Pennisetum cenchroides.
1. A portion of a spike; 2 and 3. spikelets with their involucels; 4. a bristle; 5, 6, 7 and 8. the first, second, third and the fourth glume, respectively; 7a and 8a. palea of the third and the fourth glume, respectively; 9. ovary and stamens.

[Pg 119]

Spikelets are about 1/5 inch long oblong-lanceolate, one to three in a spike and sessile.

There are four *glumes* in a spikelet. The *first glume* is small, hyaline, ovate-lanceolate, acute, nerveless or sometimes 1-nerved. The *second glume* is a little longer than the first, ovate, acute, about half of the third glume, hyaline, 1 to 3-nerved. The *third glume* is ovate-lanceolate, acuminate, generally 5- to 7-nerved, paleate, usually male; *palea* is lanceolate, equal to or slightly shorter than the glume. The *fourth glume* is as long as the third with a broad hyaline margin, 5-nerved paleate; palea as long as the glume. *Anthers* are three, yellow, *stigmas* white, feathery and the styles shortly united at the base. *Lodicules* are not present.

This is the famous Kolakattai grass (Tamil) of the Coimbatore District and it grows in all kinds of soil and is capable of growing even when the soil is dry. It is readily eaten by cattle, sheep, goat and when once established is not easily killed out even by prolonged droughts. It is in flower in June, November and December.

Distribution.—Fairly common in South India and Western India. Said to occur in Tropical Africa also.

There is a variety of this grass named *echinoides*. This differs from the type in the following respects—the inner bristles are united very much above the base and much thickened and stiffer than in the type. (See fig. 116-3)

[Pg 120]

10. Cenchrus, *L.*

The inflorescences are spike-like racemes, consisting of involucellate clusters of shortly pedicelled spikelets jointed on a simple rachis. The involucel consists of hardened spike-like bristles connate at the base into a short coriaceous cup, which is surrounded by erect or squarrose bristles. Spikelets one to three in each involucel, persistent, 1- to 2-flowered, with three or four glumes. The first glume is very small or absent. The second and the third glumes are subequal 5- to 7-nerved. The third glume is longer than the second with male flower or not, paleate. The fourth glume is coriaceous, with a bisexual or female flower. Lodicules are two. Stamens are three. Styles are long, free or connate below. Grain is broad, oblong and compressed.

KEY TO THE SPECIES

- Base of involucel rounded; inner bristles shorter, erect, not ciliateand connate at base. 1. C. biflorus.
- Base of involucel turbinate, inner bristles longer, spreading and spinescent, ciliate at base 2. C. catharticus.

[Pg 121]

Cenchrus biflorus, *Roxb.*

This is an annual with erect simple stems, 6 to 24 inches long.

The *leaf-sheath* is glabrous or nearly so, with hairs at the mouth.

The *leaf-blade* is linear-lanceolate, finely acuminate, glabrous or hairy, 3 to 10 inches long and 1/8 to 3/8 inch broad.

The *inflorescence* is a solitary cylindric raceme of involucels, 2 to 4 inches long, enclosed in the uppermost leaf-sheath; the *rachis* is flexuous, angular and smooth. *Involucels* usually with two, rarely three spikelets, loosely imbricate, rounded at the base; the inner bristles are erect, dorsally flat, subulate-lanceolate, puberulous and with thickened margins, about 1/8 inch long. The outer are shorter than the inner, glabrous, erect or subsquarrose and as long as the sessile spikelets.

The *spikelets* are about 1/6 inch long, sub-globose, with four *glumes*. The *first glume* is about 1/10 inch long, ovate-acuminate, very thin, hyaline, nerveless or rarely 1-nerved. The *second glume* is broadly ovate, 1/6 inch long, hyaline, acute, 1-nerved. The *third glume* is slightly longer than the second, oblong-ovate, apiculate, 5-nerved and paleate; *palea* 1/8 inch obtuse. The *fourth glume* is as long as the third, ovate, obtuse, paleate. *Anthers* are three. *Styles* free almost to the base. The grain is 1/12 by 1/16 inch orbicular oblong, compressed, smooth and pale brown.

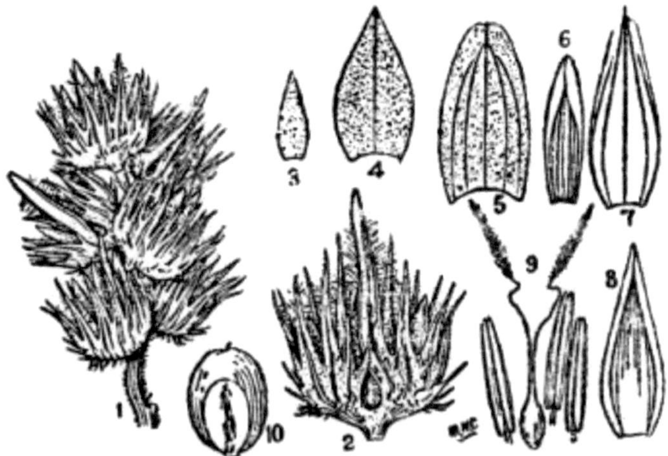

Fig. 117.—Cenchrus biflorus.
1. A portion of the raceme; 2. an involucel; 3, 4, 5 and 7. the first,

second, third and the fourth glume respectively; 6 and 8. palea of the third and the fourth glumes; 9. the ovary and stamens; 10 grain.

This grass is not so widely distributed as *Cenchrus catharticus*. It is confined to some East Coast districts.

Distribution.—The Punjab, Gangetic plain, Concan, Sind and Coromandel. Also said to occur in Africa and Arabia.

[Pg 122]

Cenchrus catharticus, *Delile.*

A tufted annual grass with geniculately ascending stems, branching at the base.

The *leaf-sheath* is glabrous and somewhat inflated. The *ligule* is a fringe of hairs. *Nodes* are glabrous.

The *leaf-blade* is linear-lanceolate, finely acuminate, 1 to 4 inches long and 1/8 to 1/4 inch broad.

The *inflorescence* is usually enclosed in the leaf-sheath, 1 to 6 inches long; the *rachis* is flexuous, angular and glabrous. The involucels are 1/4 to 1/2 inch across, turbinate or truncate at base with an outer, shorter and inner longer series of hard, sharp, pungent spines; the inner subulate, dorsally deeply grooved, very much longer than the spikelets; margins ciliate to about half the distance from the base, and the upper half covered with very short, sharp and stiff, reflexed hairs; the outer are shorter than the spikelets, spreading or erect, glabrous or nearly so and covered with reflexed hairs.

The *spikelets* are usually one to two and rarely three in an involucel and each one has four *glumes*. The *first glume* is lanceolate and nerveless or ovate-lanceolate and 1-nerved, half as long as the third glume, hyaline and acute. The *second glume* is about 1/6 inch long, ovate, acute, membranous, 5-nerved. The *third glume* is similar to the second, paleate; *palea* is lanceolate and short. The *fourth glume* is as long as the third, cuspidately acuminate, membranous, 5-nerved and paleate; *palea* is ovate, as long as the glume. *Stamens* are three. *Styles* are free and long with plumose stigmas. The grain is ovoid-oblong, brown and compressed.

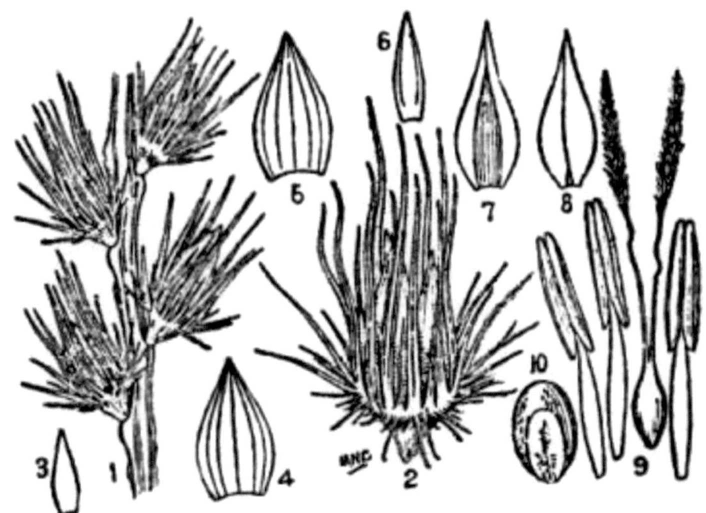

Fig. 118.—Cenchrus catharticus.
1. A portion of the spike; 2. an involucel with two spikelets; 3, 4, 5 and 7. the first, second, third and the fourth glume respectively; 6 and 8. palea of the third and the fourth glume respectively; 9. the ovary and stamens; 10. grain.

This grass is more common than *C. biflorus* and is found on the East Coast districts in open sandy places.

Distribution.—Nellore, Bellary, the Punjab and the Gangetic Plains.

[Pg 123]

CHAPTER VII.
TRIBES II AND III—ORYZEÆ AND ZOYSIEÆ.

Oryzeæ is a very small tribe with a few genera, which usually flourish in marshes. The spikelets are in panicles, 1-flowered and the flower is usually perfect. In Oryza there are three glumes, the first two being very minute, and there is only a single glume in Leersia and Hygrorhiza. There are usually six stamens. The palea becomes firm in texture like the glume instead of remaining hyaline, and so it is often mistaken for a glume. The spikelets are jointed on their pedicels and fall away from them.

- Not floating; spikelet not awned 11. Leersia.
- Floating; spikelets awned 12. Hygrorhiza.

Zoysieæ is another small tribe with half a dozen genera. The inflorescence is either a spike-like raceme or a spiciform panicle. The spikelets are solitary in Perotis, binate in Tragus and grouped in Trachys. There is usually a complete flower in a spikelet and the glumes are membranous. Mature spikelets are deciduous with their pedicels singly in Perotis and in clusters in others.

- Spikelets fascicled unilaterally on a broad rachis, 4-glumed, glumes not echinate 13. Trachys.
- Spikelets binate and all round the rachis, 3-glumed, glumes echinate 14. Tragus.
- Spikelets single, awned and 3-glumed 15. Perotis.

[Pg 124]

11. Leersia, *Sw.*

These are tall perennial marsh grasses. The inflorescence is usually a more or less contracted panicle with very slender branches. The spikelets are compressed and consist of only one glume bearing a perfect flower. The solitary flowering glume is chartaceous, awnless, 3- to 5-nerved, the lateral nerves forming the thickened margin of the glume. The palea is narrow, linear-lanceolate, as long as the glume, 3-nerved, rigid, dorsally ciliate, and with hyaline margins. Lodicules are two. Stamens are usually six in number. Styles are short, with plumose stigmas and laterally exserted. Grain is ovoid or oblong, compressed, free within the glume and its palea.

[Pg 125]

Leersia hexandra, *Sw.*

This is a slender perennial marsh-grass with stems rooting in the mud and with flexuous floating branches, sending up erect or ascending, weak and slender leafy branches, 2 to 4 feet high.

Fig. 119. — Leersia hexandra.
1. Erect branch; 2 and 3. bits of leaves with ligules; 4 and 5. spike-lets; 6. ovary and lodicules.

The *leaf-sheath* is smooth, glabrous, with eciliate margins. The *ligule* is a short obliquely truncate or two-lobed membrane. *Nodes* are hairy with deflexed hairs.

The *leaf-blade* is flat, narrow, linear, tapering to a fine point, suberect and rather rigid, glabrous and with a narrow base, varying in length from 3 to 10 inches and 1/8 to 1/3 inch in breadth.

The *inflorescence* is an oblong laxly branched, narrow pedunculate panicle, 2 to 4 inches long.

The *spikelets* are all 1-flowered and 1-glumed, articulate on the pedicels above the rudimentary glumes, strongly laterally compressed. The *glume* is about 1/6 inch long, ovate-oblong, somewhat boat-shaped, acute and shortly mucronate, strongly keeled, ciliate on the keel and margins, 5-nerved, the lateral nerves forming a thickened margin; *palea* is as long as the glume, linear-lanceolate, subacute, rigid with membranous margins. *Stamens* are six and there are two small *lodicules*. The first two glumes are reduced to an obscure hyaline rim.

This marsh-grass is found in marshy places such as ditches and channels in paddy fields, ponds and tanks.

Distribution.—It is found all over India and Ceylon; also in Africa, America and Australia.

[Pg 126]

12. Hygrorhiza, *Nees.*

These are floating glabrous grasses with stems diffusely branching and profusely rooting at the nodes. The inflorescence is a panicle. The spikelets are 1-flowered, with a solitary flowering glume only. The flowering glume is awned, strongly 5-nerved, nerves scabrid and ciliate, the lateral nerves being marginal. Palea is 3-nerved, narrow acuminate with a ciliate keel. Lodicules are suborbicular. There are six stamens with long slender anthers. Styles are free with plumose stigmas, laterally exserted. Grain is oblong, narrowed at the base, obtuse, free within the glume and its palea.

[Pg 127]

190

Hygrorhiza aristata, *Nees*.

This is a floating aquatic grass. Stems are spongy, branching diffusely, 1 foot long, with feathery whorled roots in dense masses at the nodes; branches are short, erect and leafy.

The *leaf-sheath* is smooth, inflated, compressed, with ciliate margins. The *ligule* is a narrow membrane. *Nodes* have whorls of roots.

The *leaf-blade* is linear or ovate-lanceolate, obtuse, glabrous, glaucous beneath, base rounded or subcordate, 1 to 3 inches long and 1/2 to 3/4 inch broad.

The *inflorescence* is a panicle, 2 inches long and broad, somewhat triangular in outline; the *rachis* and the branches are stiff, slender and smooth, the lower branches are a little deflexed.

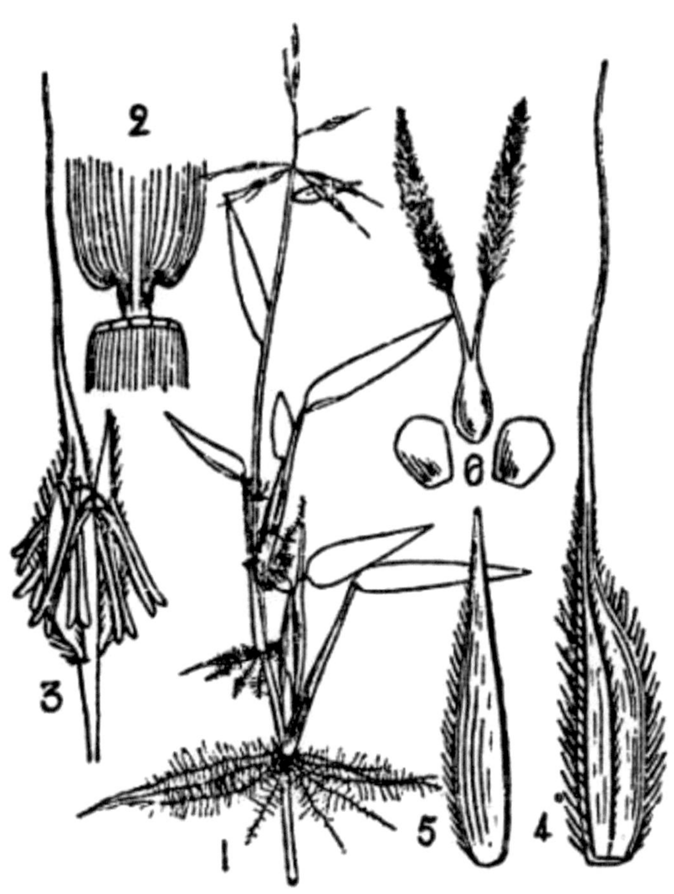

Fig. 120.—Hygrorhiza aristata.
1. Branch; 2. part of a leaf with ligule; 3. spikelet; 4 and 5. glume and
its palea; 6. lodicules and ovary.

The *spikelets* are very narrow, sessile or pedicellate, articulated on
the pedicel, 1-flowered and 1-glumed. The *glume* is about 3/8 inch
long (excluding the awn) and the awn is as long as the glume or
slightly longer, lanceolate, with five strong nerves and the lateral
ones forming thickened margins; the palea is as long as the glume.
Stamens are six and *lodicules* two.

Found in ponds and tanks.

Distribution.—All over India and Ceylon.

[Pg 128]

13. Trachys, *Pers.*

These are softly, villous, diffuse annual grasses. The inflorescence consists of usually two (rarely three) divaricating spikes on a long peduncle. The rachis is herbaceous, broad flexuous, jointed and bearing at each joint a solitary globose cluster of two or three perfect 1-flowered glabrous spikelets surrounded by many short spinescent glumes of imperfect ones. The perfect spikelets are 4-glumed and the glumes are very unequal. The first glume is minute, tooth-like, nerveless. The second glume is long, linear-lanceolate, membranous, very acute, strongly 3- to 5-nerved. The third glume is the largest, obliquely ovate, or obovate-oblong, cuspidately acuminate, rigidly coriaceous, 9- to many-nerved, paleate or not, empty. The fourth glume is shorter and narrower than the lower one, linear-oblong, acuminate, chartaceous, smooth, dorsally convex, with incurved margins, bearing a bisexual flower, paleate, palea is hyaline as long as the glume, and the margins are inflexed below the middle. Lodicules are very minute or wanting. There are three stamens. The styles are very long with slender stigmas, exserted at the top of the glume. The grain is oblong, compressed, free within the glume and its palea.

[Pg 129]

Trachys mucronata, *Pers.*

This is a diffusely branching, softly villous annual grass. The stems are many from the root, 16 to 18 inches long, ascending or decumbent and prostrate, leafy, glabrous, rooting freely at the lower nodes, especially when procumbent.

The *leaf-sheaths* are loose, inflated, hairy or rarely glabrous. The *ligule* is a thin membrane, or a ridge of fine closely set hairs. *Nodes* are villous.

The *leaf-blade* is linear-lanceolate to ovate-lanceolate acuminate, flaccid, softly villous on both the surfaces, margins often crisped, base rounded, 2 to 6 inches by 1/4 to 1/2 inch.

The *inflorescence* consists of a long or short, slender, shining peduncle bearing two or three rigid, flattened, flexuous, jointed spikes, the rachis is broad, herbaceous, with a flat, broad, closely nerved wing on both the sides and with a distinct flat midrib and jointed, each joint bears on the under surface at the articulation, a solitary, globose cluster of two to three perfect 1-flowered glabrous spikelets surrounded by many short spinescent glumes of imperfect ones. The spikes vary in length from 1 to 2 inches and in breadth from 1/10 to 1/6 inch and are glabrous.

The clusters of *spikelets* are about 1/4 inch in diameter, often partially sunk, in a concavity of the rachis; the perfect spikelets are 1/5 to 1/4 inch long and the imperfect are shorter.

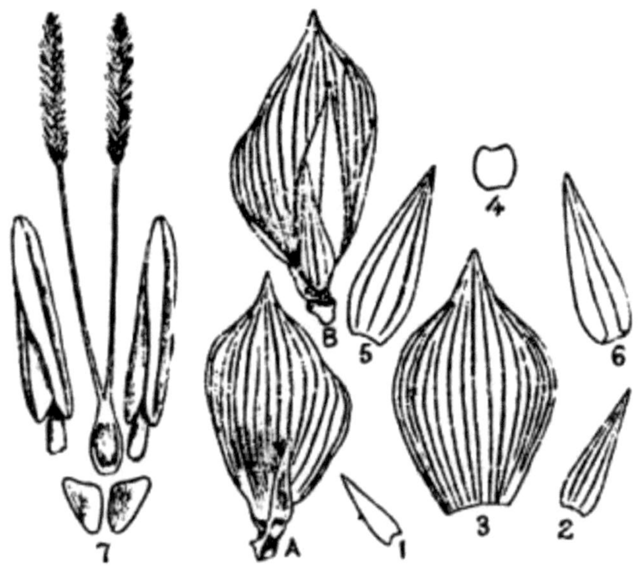

Fig. 121.—Trachys mucronata.
A and B. The spikelets; 1, 2 and 3. the first, second and the third

glume, respectively; 4. palea of the third glume; 5 and 6. the fourth glume and its palea; 7. lodicules, ovary and stamens.

In the perfect spikelet there are four very unequal glumes. The *first glume* is minute, tooth-like, triangular or lanceolate, acute, nerveless, 1/16 to 1/12 inch long. The *second glume* is elongate, linear-lanceolate, acute, sometimes ciliate below the middle, membranous, narrower than the third glume, hyaline, strongly 3-nerved, 1/16 by 1/6 inch. The *third glume* is 1/5 by 1/8 inch the largest in the spikelet, broadly and obliquely ovate or obovate, [Pg 130] cuspidately acute, with nine to many green nerves, paleate; the *palea* is very small, about 1/20 inch long, oblong, hyaline and rigidly coriaceous. The *fourth glume* is much narrower and shorter than the third glume, linear oblong, acuminate, chartaceous, smooth, dorsally convex, with incurved margins, bisexual and paleate; the palea is as long as the glume, acuminate, hyaline, the margins inflexed below the middle, ovate, acute. *Lodicules* are minute or absent. *Stamens* are three with linear anthers. *Styles* are very long with slender stigmas. The grain is oblong, compressed.

This grass grows abundantly in cultivated dry fields and in the sand near the sea-shore and it is easily recognized by the clusters of spikelets in the spike.

Distribution. — The Deccan Peninsula — both in the interior and on the sea coast.

[Pg 131]

14. Tragus, *Haller.*

These are annual or perennial grasses, with erect or prostrate stems. Inflorescence is a spiciform raceme, bearing the spikelets in clusters of 2 to 4. The spikelets are 1-flowered and usually with two glumes. Sometimes a very minute hyaline lower glume is present. The first glume is thickly coriaceous, 5-ribbed, oblong-lanceolate, and ribs with long recurved spines. The second glume is oblong or oblong-lanceolate, apiculate, chartaceous, 3-nerved and with a perfect flower; palea is as long as the glume, 2-nerved. Lodicules are broad, cuneate and fleshy. There are three stamens. Styles are slender and distinct, with narrow stigmas exserted from the top of the glume. Grain is oblong to ellipsoidal free within the glume and its palea.

[Pg 132]

Fig. 122.—Tragus racemosus.

[Pg 133]

Tragus Racemosus, *Scop.*

This plant is a perennial with tufted prostrate or erect stems, rooting at the nodes freely and densely leafy. The flowering branches are erect or geniculately ascending and varies from a few inches to about a foot.

The *leaf-sheath* is short, pale, glabrous, somewhat compressed, striate, equitant below and upper are longer, terete and green. The *ligule* is only a ridge of short, fine hairs. *Nodes* are glabrous.

The *leaf-blade* is convolute when young, ovate or ovate-lanceolate, variable from 1/4 to 2 inches long and 1/10 to 1/6 inch wide, acuminate, flat or somewhat wavy, glabrous on both the surfaces, rigidly pungent, densely crowded and distichously imbricate in the lower part of the stem, base is amplexicaul, and the margin is distantly serrate and rigidly ciliate.

The *inflorescence* is a spike-like terminal panicle varying in length from 3/4 to 2 inches; the *rachis* is wavy, slender, angular or grooved, pubescent, the peduncle is striate, pubescent and enclosed by the leaf-sheath.

The *spikelets* are arranged in groups of two, facing each other and appearing like a single spikelet with two equal echinate glumes, sessile, or obscurely pedicelled on very short, tumid, pubescent branches.

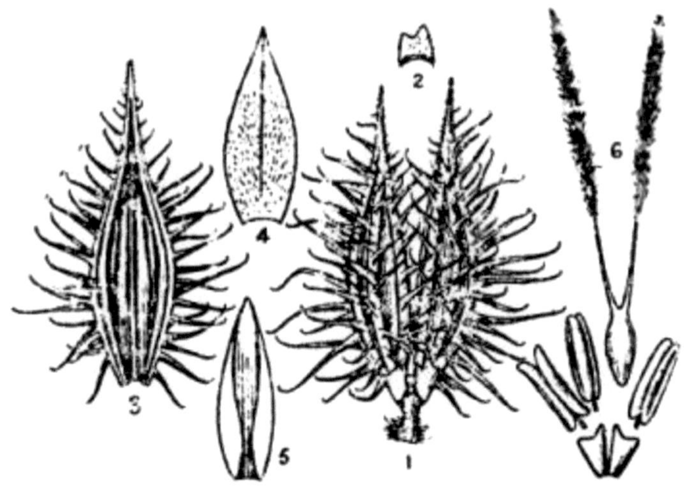

Fig. 123.—Tragus racemosus.
1. A pair of spikelets; 2, 3 and 4. the first, second and the third glume, respectively; 5. palea of the third glume; 6. ovary, anthers and lodicules.

There are two (rarely three) *glumes* in the spikelet. The *first glume* is very minute, hyaline, obtuse and it is very often not present. The *second glume* is about 1/8 inch, ovate-lanceolate, acuminate, strongly 3-ribbed with rows of stout, spreading hooked spines along the ribs and encloses a single floret. The margins of this glume are membranous and somewhat scaberulous. The *third glume* is about 1/12 inch, oblong lanceolate, membranous minutely hairy, 3-nerved and finely pointed at the apex; the *palea* is as long as the glume, hyaline, 2-nerved, lanceolate, subacute and irregularly toothed at the apex. *Stamens* are three, with slender [Pg 134] filaments, anthers are short, broad and pale yellow. The style branches are pale and feathery. *Lodicules* are two, fleshy and cuneate or subquadrate. The grain is free inside the glume and the palea, linear oblong, slightly compressed and pale brown, the embryo occupies about 1/3 the length of the grain.

This is one of the commonest grasses growing everywhere in tufts with usually prostrate branches. In some situations the branches are erect.

Distribution.—Plains of India throughout and Ceylon. It is found in all the warm regions of the world.

[Pg 135]

15. Perotis, *Ait.*

These are slender annual or perennial grasses with short broad leaves. Inflorescence is a spike or spiciform raceme. The spikelets are 1-flowered, sessile or shortly pedicelled and jointed. There are three glumes in the spikelet. The first and the second glumes are empty, subequal, narrowly linear with a strong midrib which is produced into a long capillary awn. The third glume is very small, hyaline, lanceolate, acute, 1-nerved and with a perfect flower; palea is small, narrow, hyaline and nerveless. Stamens are three with short anthers. Styles are short and united at the base with very short stigmas. The grain is long and narrow, longer than the flowering glume.

[Pg 136]

Fig. 124.—Perotis latifolia.

[Pg 137]

Perotis latifolia, *Ait.*

This grass is an annual with slender leafy stems, branching at the base, prostrate at first and then geniculately ascending, terminating in inflorescences and varying in length from 3 to 15 inches.

The *leaf-sheaths* are glabrous, usually all short except the one next to the inflorescence which is two or three times as long as the lower sheaths. The *nodes* are purple and glabrous.

The *leaf-blade* is short, 1 to 1-1/4 inches long, ovate or lanceolate, cordate at base, acute and glabrous on both the surfaces; the margin is minutely serrate, rigidly ciliate and with a very narrow hyaline border.

The *inflorescence* is a slender, crinite, spike-like raceme, 1 to 8 inches long, with a finely scabrid main *rachis*.

The *spikelets* are narrow linear 1/12 to 1/8 inch or longer, purple, shortly pedicelled and 1-flowered, pedicels are short with a hyaline swelling on the upper side at the base.

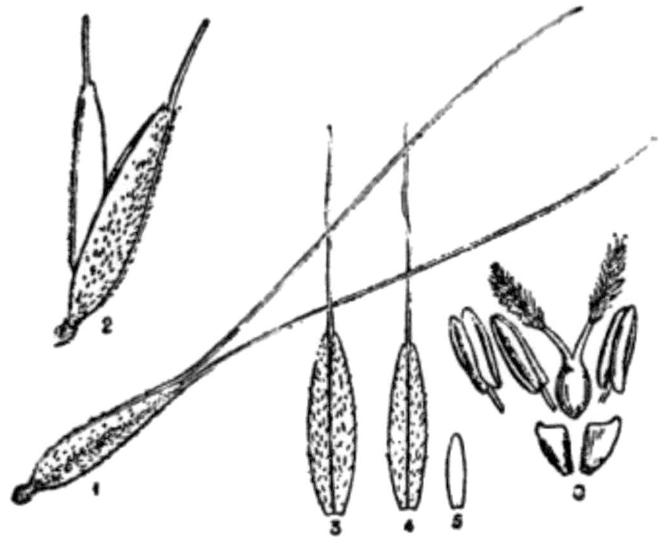

Fig. 125.—Perotis latifolia.
1 and 2. Spikelets; 3, 4 and 5, the first, second and the third glume, respectively; 6. ovary, stamens and lodicules.

There are three *glumes*. The *first* and the *second glumes* are empty, narrow-linear, purple, scabrid, 1-nerved and awned; awns are capillary, varying in length from 1/3 to 1/2 inch. The *third glume* is very minute with very small palea. There are three *stamens* and two small *lodicules*. *Styles* are somewhat shorter. The grain is long and cylindric.

This grass grows in open waste places and in dry fields all over the Presidency.

Distribution.—Throughout India.

[Pg 138]

CHAPTER VIII.
TRIBE IV — ANDROPOGONEÆ.

Andropogoneæ is a very large tribe with about thirty genera. It is very well represented in South India and some genera are of very wide distribution.

The spikelets are usually arranged in pairs at each joint, one sessile and the other stalked. The spikelets may all be similar as in Imperata or they may be different as in Ischæmum and Andropogon. There may be only one flower in the spikelet as in Eremochloa and Saccharum or two as in Ischæmum and Apocopis. In the genera Polytoca and Coix the spikelets are unisexual and the male and female spikelets are found in the same inflorescence, the female being below and the male being continuous with it. The spikelet nearly always consists of four glumes, the first or the first and the second being firmer and coriaceous or chartaceous. The flowering glumes are always shorter than the empty glumes, and are hyaline. The fourth glume is often awned or reduced to an awn.

The main rachis of the inflorescence is usually jointed at the base. In addition to this the rachis may be jointed all along its length, so as to become separated into distinct joints when mature as in Rottboellia, Saccharum and Andropogon, or it may be continuous as in Imperata. The pedicels of spikelets and the lower portions of the rachilla of the spikelets may have long hairs.

Sub. Tribe 1. **Maydeæ.**

- The spikelets are all unisexual, spicate, the male and female spikelets are dissimilar, and are on the same or on different spikes.
    - o Fruiting spikelets enclosed in a stony nut-like polished bract 16. Coix.
    - o Fruiting spikelets with the first glume forming a crustaceous nut-like envelope to other glumes and grain 17. Polytoca.

Sub. Tribe 2. **Sacchareæ.**

- The spikelets are all similar, in compound racemes or panicles; the first glume not sunk in the hollow of the rachis. Spikelets are 1-flowered.
  - Rachis not fragile; spikelets in cylindrical silvery thyrsus 18. Imperata.
  - Rachis fragile; spikelets in open very much branched silky panicles 19. Saccharum.

Sub. Tribe 3. **Ischemeæ.**

- Spikelets many, dissimilar, in solitary, digitate or fascicled racemes or spikes; first glume not sunk in the hollow of the rachis. [Pg 139]
  - Margins of the first glume of the sessile spikelet inflexed.
    - Spikes rarely solitary; spikelets binate, 2-flowered and awned 20. Ischæmum.
    - Spikes solitary; spikelets 1-flowered; first glume of the sessile spikelet pectinate 21. Eremochloa.
  - Margins of the first glume of the sessile spikelet not inflexed.
    - Spikes solitary or binate; spikelets 1- to 2-flowered, diandrous; first glume broad and truncate 22. Apocopis.
    - Spikes 2 or more; spikelets binate, upper alone awned 23. Lophopogon.

Sub. Tribe 4. **Apludeæ.**

- Spikelets three on an inarticulate rachis 24. Apluda.

Sub. Tribe 5. **Rottboellieæ.**

- Spikelets similar or dissimilar, 1- to 2-flowered, solitary, 2- or rarely 3-nate on the internodes of an articulated spike or raceme, not awned; the first glume is not keeled, sunk in a cavity of joints of the rachis; sessile spikelets 4-glumed.

- o Sessile spikelets single; first glume flat 25. Rottboellia.
- o Sessile spikelets geminate in all except the uppermost joints 26. Mnesithea.
- o Sessile spikelets binate; first glume globose, pitted 27. Manisuris.

Sub. Tribe 6. **Eu-Andropogoneæ.**

- Spikelets are dissimilar, 1-flowered, 2-(rarely) 3-nate on the whorled articulate branches of simple or compound racemes or panicles; glumes four, first glume not keeled, fourth glume usually awned.
  - o Spikelets binate below and 3-nate at the top on a spicate or panicled inflorescence 28. Andropogon.
- Spikelets in two superposed series. Upper series of one or more sessile bisexual or female spikelets with one terminal pedicelled male spikelet.
  - o Rachis jointed above the involucral spikelets 29. Anthistiria.
  - o Rachis jointed below the involucral spikelets 30. Iseilema.

[Pg 140]

16. Coix, *L.*

These are tall monoecious annual or perennial grasses. The inflorescences are terminal or axillary spiciform racemes. The lowest-spikelet in the raceme is female and this is enclosed in a bract which at length becomes hardened, polished and nut-like and the other spikelets above it are male. The male spikelets are 2- to 3-nate at each node of the rachis, 1 sessile and 1 or 2 pedicelled, lanceolate and 4-glumed. The first and the second glumes are subequal and empty, and the first glume is winged along the inflated margins. The third and the fourth glumes are hyaline, with three stamens or empty. The female spikelet is ovoid acuminate and has four glumes. The first glume is chartaceous and the others are thin and gradually smaller. The grain is orbicular, ventrally furrowed and enclosed by the polished hard bract.

[Pg 141]

Coix lachryma-jobi, *L.*

This is a tall monoecious leafy annual (rarely perennial) grass with stout, smooth, polished, freely branching stems rooting at the lower nodes and varying in length from 3 to 5 feet or more.

The *leaf-sheath* is long, usually smooth but occasionally with scattered tubercle-based hairs. The *ligule* is a narrow membrane. The *nodes* are glabrous.

The *leaf-blade* is long, flat, narrow or broad, acuminate, cordate at base, with a stout midrib and many slender veins on both sides, usually glabrous on both sides though occasionally with scattered hairs, and with spinulosely serrate margins, varying from 4 to 18 inches in length and 1/3 to 2 inches in breadth.

The *inflorescence* consists of nodding or drooping spiciform racemes, 1 to 1-1/2 inches long, terminating the branches. The racemes consist of many male spikelets with one (rarely two) female spikelets at the base; the rachis is stout above, and the part within the bract enclosing the female spikelet is slender.

The *male spikelets* are imbricating, 2 or 3 at a node of the rachis, one sessile and one or two pedicelled, dorsally compressed, articulate at the base and persistent, very variable in size, 3/8 to 3/4 inch. There are four *glumes* in the spikelet. The *first glume* is oblong-lanceolate, chartaceous, 3/5 inch long, acute, many-nerved, concave with inflexed margins bearing narrow green many-veined wings. The *second glume* is similar to the first, but thinner and without the wings, 5- to 9- or rarely 11-nerved. The *third glume* is oblong-lanceolate, hyaline, faintly 3- to 5-nerved, paleate and with three stamens. The *fourth glume* is similar to the third, paleate with or without stamens.

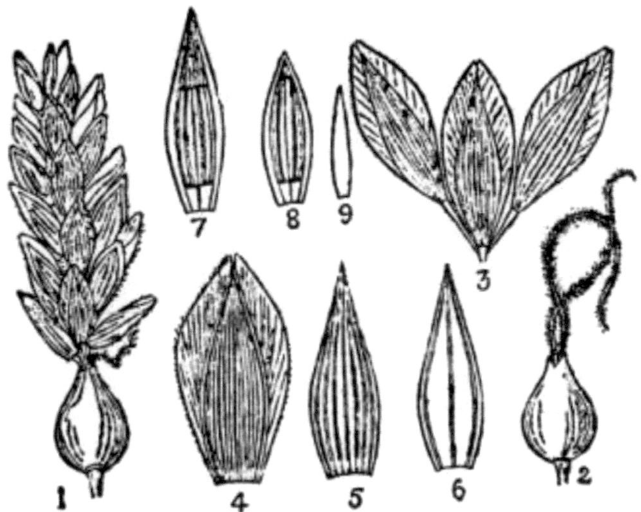

Fig. 126.—Coix Lachryma-Jobi.

1. Inflorescence; 2. the female spikelet; 3. male spikelets; 4, 5, 6 and 8. the first, second, third and the fourth glume, respectively, of a male spikelet; 7 and 9. palea of the third and the fourth glumes, respectively.]

The *female spikelet* is enclosed by a closed bract which finally becomes hardened, and there are four *glumes* in the spikelet. The [Pg 142] *first* and the *second glumes* are chartaceous. The *third* and the *fourth* are hyaline, the former being empty and the latter with an ovary. *Lodicules* are not present. The *ovary* is ovoid with very long capillary styles. The grain is orbicular, compressed, channelled at the back and enclosed within the stony, hardened and polished bract.

This grass usually grows in paddy fields. There are two distinct varieties—one a fairly tall one annual and the other a very tall (5 to 10 feet) perennial one. The racemes of the latter are longer and drooping, the male spikelets are in threes and the wings of the first glume are usually broader than in the other form. This species is easily recognized by the polished bract enclosing the female spikelet.

207

Distribution. — Throughout India.

[Pg 143]

17. Polytoca, *Br.*

These are tall monoecious annual or perennial grasses. Inflorescences consist of spiciform racemes with spathaceous bracts; rachis is jointed. Racemes may all be male or with one or two female spikelets at the base. Male spikelets are geminate, one sessile and one pedicelled, 2-flowered or imperfect, and with four glumes, which are subequal. The first glume is membranous, many-nerved, shallowly concave and with a narrow membranous margin. The second glume is narrower, ovate, acute, 5- to 9-nerved. The third glume is membranous, oblong, acute, 3- to 5-nerved, paleate and with three stamens. The fourth glume is very slender, hyaline, linear, paleate with three stamens or empty. Female spikelets are broadly oblong, 1-flowered and with four glumes. The first glume is thick, coriaceous and closely embraces the rachis of the spike by its involute margin and the other glumes are within. The second glume is oblong, many-nerved. The third is narrowly oblong, 3- to 5-nerved, empty. The fourth glume is very narrow, truncate, 3-nerved, paleate. Styles are very long with slender stigmas. Grain is small, fusiform, terete and enclosed in the nut-like polished and hardened first glume.

[Pg 144]

Polytoca barbata, *Stapf.*

This is an erect, tall, stout, freely branching, leafy, monoecious perennial grass. The stems are terete, 3 to 6 feet high.

The *leaf-sheaths* are long, glabrous, or with scattered tubercle-based bristly hairs. The *ligule* is a narrow membrane. The *nodes* have a ring of soft long hairs.

The *leaf-blades* are long, flat, linear, acuminate, with a stout midrib and thickened serrate margins, scabrid above and sometimes with a few tubercle-based hairs, 10 inches to 2 feet long and 1/4 to 3/4 inch broad.

208

The *inflorescence* consists of paniculate spike-like racemes terminating the branches and at first enclosed in spathiform bracts, the lower and outer spathiform bracts are one inch or more in length with a long awn at the tip, and the inner proper sheaths are oblong, awned and about 1/2 inch long. The raceme consists of one or more female spikelets at the base and a number of male spikelets above, appearing as if sessile on the top of the female spikelet, but really articulate with the internode below it which is enclosed by the first glume of the female spikelet.

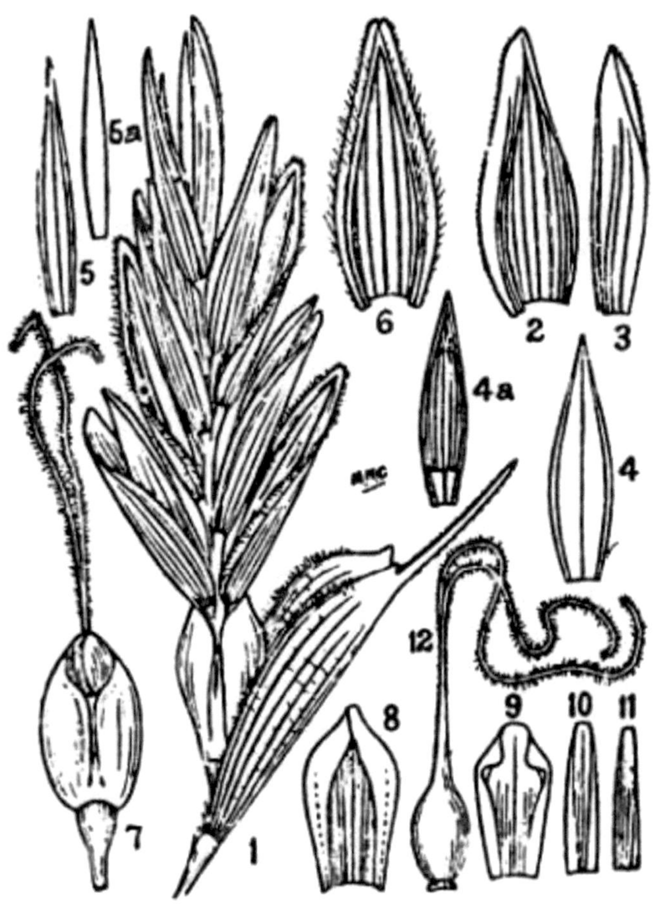

Fig. 127.—Polytoca barbata.
1. Inflorescence; 2, 3, 4 and 5. the first, second, third and the fourth glume, respectively, of the male spikelet; 4a and 5a. palea of the third and the fourth glume, respectively; 6. the first glume of the sessile spikelet; 7. female spikelet; 8, 9 and 10. the second, third and the fourth glume, respectively; 11. palea of the fourth glume; 12. ovary.

The *male spikelets* are solitary, or binate and then one sessile and one pedicelled, 2-flowered, reaching 3/8 inch in length and consist of four *glumes* each. The *first glume* is concave, ovate, acute, [Pg 145] pubescent, herbaceous, many-nerved and with a narrow membranous margin on one side only in the pedicelled and solitary spikelets and on both sides in the sessile spikelets. The *second glume* is narrower, dorsally compressed, ovate, acuminate, 5- to 9-nerved, laterally compressed and with a narrow wing to the keel near the apex in sessile spikelets and dorsally compressed without the keel in the pedicelled and solitary spikelets. The *third glume* is membranous, oblong, acuminate, 3- to 5-nerved, with three stamens and paleate; the *palea* is hyaline, broadly linear. The *fourth glume* is very slender, linear, hyaline, with or without stamens, paleate; *palea* is flat, narrowly linear. *Lodicules* are present and they are small. The *anthers* in the third glume are larger than those in the fourth glume.

The *female spikelet* is oblong, 1/6 inch long, 1-flowered and with four *glumes*. The *first glume* is thickly coriaceous, white, shining, closely embracing the rachis and the other glumes entire at the tip. The *second glume* is quadrately oblong, many-nerved. The *third glume* is oblong, narrower than the second, 3- to 5-nerved paleate, empty. The *palea* of the third glume is narrow, truncate. The *fourth glume* is narrow, truncate, 3-nerved, paleate; the *palea* is truncate and wrapped round the ovary. *Styles* are long and stigmas slender. *Lodicules* are not present. The grain is fusiform, terete and within the nut-like polished hardened glume.

Distribution. — In damp situations all over India.

[Pg 146]

18. Imperata, *Cyril.*

These are erect perennial grasses. The inflorescence is a spike-like panicle, with very short filiform inarticulate branches and rachises. Spikelets are binate, 1-flowered, all alike, both pedicelled, articulate at the base and hidden by the very long silky hairs arising from a small callus and from the glumes. There are four glumes. The first two glumes are membranous, lanceolate, and subequal. The third glume is shorter and smaller, hyaline. The fourth glume is still smaller and hyaline. Stamens are two, rarely one. Lodicules are not found. Styles connate below, with stigmas very long, narrow and exserted at the top of the spikelets. Grain is small and oblong.

[Pg 147]

Imperata arundinacea, *Cyril.*

This is an erect perennial grass with creeping, stoloniferous root-stocks, with aerial stems varying from 6 inches to 3 feet.

The *leaf-sheath* is loose and glabrous. The *ligule* consists of long soft hairs. The *nodes* are naked or bearded.

The *leaf-blade* is linear, flat, tapering from about the middle to-wards the top, finely acuminate, and also narrowing towards the base into the stout midrib, margins with fine long hairs at the base, 6 to 18 inches by 1/10 to 1/3 inch, scabrous above and smooth beneath.

The *panicle* is narrow, spike-like, silvery, 3 to 8 inches; branches are short and appressed and the internodes of spikes are short with the tips dilated.

The *spikelets* are 1/8 to 1/6 inch concealed by long silvery hairs of the callus and the glumes, articulate at the base; callus hairs are about twice as long as the spikelet or longer.

There are four *glumes* in the spikelet. The *first glume* is ovate-lanceolate, obtuse, with ciliate tips and long hairs at the back below the middle, rather thickened towards the base, dorsally hairy, 3- to 7-nerved, nerves not reaching the tip. The *second glume* is as long as the first, with membranous margins and with long hairs at the back, 3- to 7-nerved. The *third glume* is hyaline, less than half as long as the first and second glumes, oblong, obtuse or irregularly toothed, nerveless or 1-nerved. The *fourth glume* is slightly shorter and narrower than the third, ovate, acute, obtuse or toothed, ciliate, nerveless or faintly 1-nerved, paleate; *palea* is about half as long as the glume, quadrate, toothed or retuse, nerveless, glabrous. There are only two stamens with orange anthers. Styles are slender, long, with purple *stigmas. Lodicules* are absent. Grain is small and oblong.

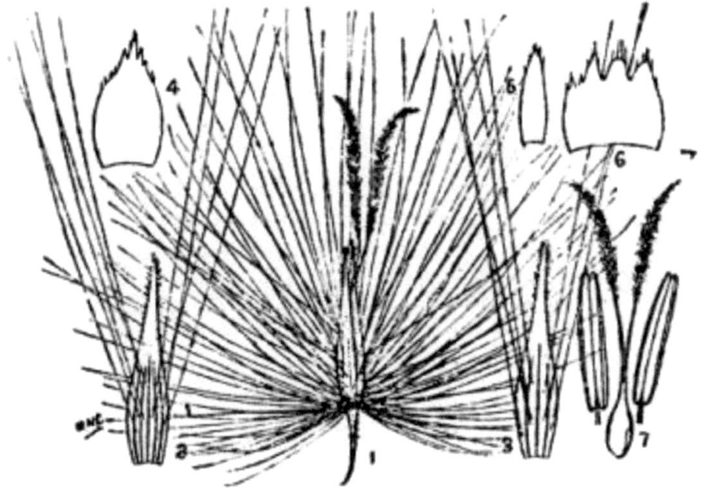

Fig. 128.—Imperata arundinacea.
1. A spikelet; 2, 3, 4, and 5. the first, second, third and the fourth glume, respectively; 6. palea of the fourth glume; 7. two stamens and the ovary.

This is fairly abundant in moist stiff soils. On account of the underground stolons this grass cannot be eradicated easily.

Distribution.—Throughout India.

[Pg 148]

19. Saccharum, *L.*

These are tall perennial grasses. Inflorescence is a much branched open panicle, branches spreading or erect, capillary and fragile. Spikelets are small, 1-flowered, binate, one sessile and the other pedicelled, the sessile spikelet is bisexual and the pedicelled is female and rarely bisexual; sessile spikelets are deciduous with the contiguous joint of the rachis and the pedicel. There are four glumes. The first glume is chartaceous, equal in length to the second, oblong or lanceolate. The second glume is concave. The third glume is hyaline, empty. The fourth glume is very small or absent.

Lodicules are present. There are three stamens. Stigmas are laterally exserted. Grain is oblong or sub-globose.

[Pg 149]

Saccharum spontaneum, *L.*

This is a tall perennial grass with a creeping root-stock bearing erect stems and occasionally decumbent or prostrate stolons. Stems vary in length from 5 to 20 feet. Branches and axillary buds grow out piercing the sheaths near the nodes.

The *leaf-sheath* is glabrous, but woolly at the mouth. The *ligule* is a distinct ovate membrane. The *nodes* are glabrous.

The *leaf-blade* is very long, narrow linear, acuminate and narrowing downwards into the stout midrib, coriaceous, glabrous and 1-1/2 to 2 feet by 1/8 to 1/4 inch.

The *panicle* is lanceolate, 8 to 24 inches, silky and the peduncle just below the panicle is softly silky, branches are whorled, three to five at a level, 2 to 4 inches long, rachis of the branches almost capillary, jointed and fragile, joints with long cilia at the back.

The *spikelets* are binate, one sessile and another pedicelled, both bisexual and alike, lanceolate, 1/8 to 1/6 inch long, callus is minute and bearded with spreading silky hairs 1/2 inch long.

214

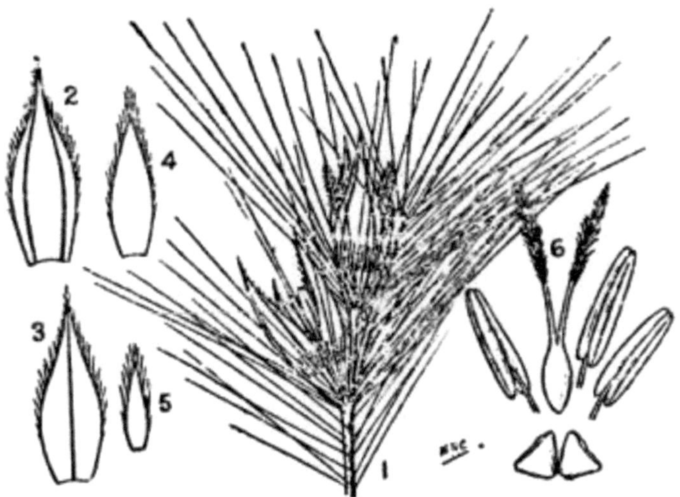

Fig. 129.—Saccharum spontaneum.
1. Two spikelets; 2, 3, 4 and 5. the first, second, third and the fourth glume, respectively; 6. ovary, stamens and lodicules.

There are four *glumes*. The *first glume* is lanceolate, subulate, acuminate, 2-nerved, flattened dorsally, coriaceous at the base and hyaline above it, and with smooth incurved margins. The second *glume* is about equal to or slightly shorter than the first, lanceolate, acuminate, 1-nerved, keeled with an opaque base; margins and keel are ciliate with fine long hairs. The *third glume* is hyaline, ovate-lanceolate, nerveless, acute, ciliate. The *fourth glume* is very slender, ciliate, acuminate, paleate; *palea* is minute, very variable. *Stamens* are three. *Lodicules* are cuneate or quadrate. The grain is very small, oblong.

Distribution.—This occurs all over India along the sides of the river.

[Pg 150]

20. Ischæmum, *L.*

The grasses of this genus are either annuals or perennials. The inflorescence consists of spikes, solitary, digitate or fascicled, articulate and fragile; the joints of the floral axis and the pedicels of the pedicelled spikelets are trigonous and hollowed ventrally. Spikelets are binate, one sessile and one pedicelled; the pedicelled spikelets are dissimilar from the sessile and both usually 2-flowered. The sessile spikelets have four glumes. The first glume is coriaceous, oblong or lanceolate, convex more or less, marginally winged above the middle, truncate or two-cuspidate at the apex and awnless. The second glume is as long as the first, coriaceous, concave, acute or obtuse, awned or not. The third glume is hyaline, deeply cleft into two lobes with an awn in the cleft, and 3-nerved, paleate; palea is linear-lanceolate enclosing either stamens and ovary or ovary alone. Lodicules are cuneate or quadrate.

KEY TO SPECIES.

- Racemes two or three; the first glume of the sessile spikelet dorsally flat, not channelled or depressed along the middle line.
  - Margin of the first glume of the sessile spikelet incurved narrowly from the base to the apex.
    - First glume of sessile spikelets with nodulose margins. 1. I. aristatum.
    - First glume of sessile spikelets closely transversely ribbed. 2. I. rugosum.
    - First glume of the sessile spikelet translucent, bicuspidate at the tip and with smooth margins. 3. I. pilosum.
  - Margin of the first glume of the sessile spikelet broadly incurved from below the middle.
    - First glume of the sessile spikelet with smooth margins, callus bearded. 4. I. ciliare.
- Raceme solitary; the first glume of the sessile spikelet deeply grooved at the back along the middle line. 5. I. laxum.

[Pg 151]

Ischæmum aristatum, *L*.

This is a perennial grass, with fairly stout, erect or somewhat decumbent, simple or branched, glabrous, leafy stems, 1 to 4 feet high.

The *leaf-sheath* is loose, glabrous and auricled. The *ligule* is a distinct membrane, broad or narrow. *Nodes* are glabrous.

The *leaf-blade* is linear-lanceolate, flat, acuminate, narrowed towards the base which may be acute, subcordate or rarely even petiolate, glabrous or sparsely hairy above and glaucous beneath, 4 to 10 inches long and 1/4 to 1 inch broad.

The *inflorescence* consists of one or two, erect, stout or slender, fragile racemes, 1 to 5 inches long.

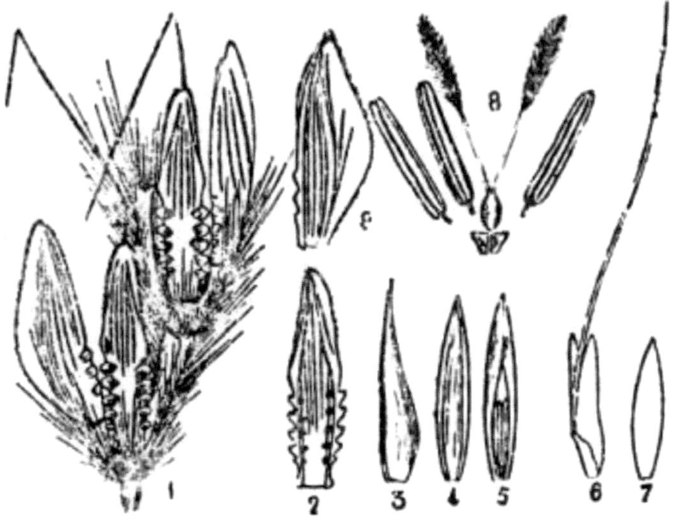

Fig. 130.—Ischæmum aristatum.
1. A portion of the raceme showing the joints, sessile and pedicelled spikelets; 2, 3, 4 and 6. the first, second, third and the fourth glume, respectively, of the sessile spikelet; 5 and 7. palea of the third and the fourth glumes, respectively; 8. ovary, stamens and lodicules; 9. first glume of the pedicelled spikelet.

The *spikelets* are 1/6 to 1/3 inch long, the sessile and the pedicelled closely pressed together, glabrous or hairy; the callus of the sessile spikelet broad and thick, with or without hairs. The *sessile spikelet* is awned and consists of four glumes. The *first glume* is 1/5 inch long or less, oblong or linear-oblong, cartilaginous below the middle, with two to four (or rarely up to six) marginal nodules on each edge, sometimes these are connected by shallow ridges, thinner above the middle, with green anastomosing veins, tip obtuse or 2-toothed, and margins narrowly incurved. The *second glume* is chartaceous, lanceolate, acuminate, 1-nerved and with a smooth rounded keel. The *third glume* is ovate-lanceolate, membranous, 1-nerved, acuminate, male or bisexual with an oblong palea. The *fourth glume* is cleft to or below the middle into lanceolate acute lobes, with a brownish red awn 1/2 inch or more long at the sinus twisted at the lower portion and straight above, paleate, usually female; *palea* is linear oblong. The *pedicelled spikelet* is as long as the sessile, inarticulate on the very thick, short pedicel which is [Pg 152] densely or sparsely hairy at the base. The *first glume* is scimitar-shaped, coriaceous, acute, with a somewhat semi-circular wing. The other *glumes* are as in sessile spikelets, but the fourth glume has no awn and may have a mucro.

This grass is a variable one. There is much variation in the breadth of the leaves and in the markings and hairiness of the spikelets. The spikelets may be glabrous or hairy and the marking in the first glume of the sessile spikelets varies in the matter of marginal nodules—it may have mere shallow notches or deep well-formed nodules and there may be transverse ridges or they may be absent. This grass is abundant on the West Coast and rare in the East Coast.

Distribution.—Throughout the plains and lower hills of India and Ceylon.

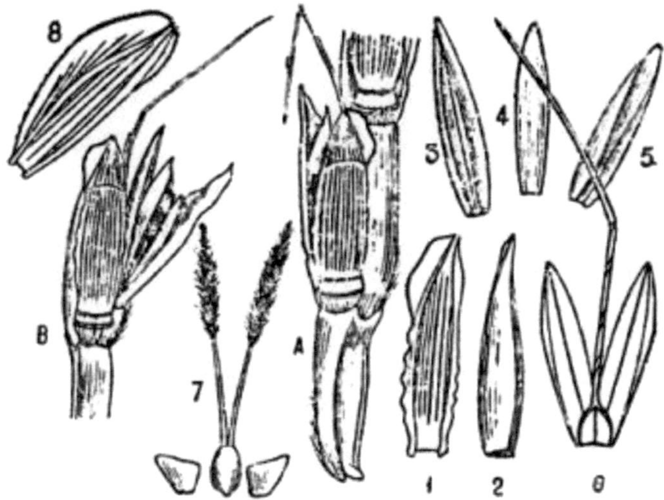

Fig. 131.—Ischæmum aristatum.
A. A portion of the raceme showing the joints; B. a sessile and a pedicelled spikelet. 1, 2, 3 and 6. the first, second, third and the fourth glume, respectively; 4 and 5. palea of the third and the fourth glume; 7. ovary and lodicules; 8. first glume of the pedicelled spikelets.

[Pg 153]

Ischæmum rugosum, *Salisb.*

This is an erect annual grass with tufted, leafy, compressed stems varying in length from 10 inches to 2 feet.

The *leaf-sheath* is glabrous, loose and compressed, with a membranous auricle confluent with the truncate *ligule. Nodes* usually glabrous but sometimes also puberulous.

The *leaf-blade* is narrow, linear-lanceolate, flat, base contracted, flaccid, acuminate, rounded at the base, glabrous or sparsely hairy on both the surfaces; the topmost leaf is often reduced to an inflated sheath enclosing the inflorescence partially.

Fig. 132.—Ischæmum rugosum.
1. A part of the raceme showing sessile spikelets with reduced pedicelled spikelets; 2. a sessile spikelet and a well developed pedicelled spikelet; 3. a reduced pedicelled spikelet; 4, 5, 6 and 8. the first, second, third and the fourth glume of the sessile spikelet; 7 and 9. palea of the third and fourth glumes of the sessile spikelet; 10. ovary.

The racemes are usually two, erect, fragile, 1 to 3 inches long with a slight thickening of the peduncle below the inflorescence; the joints are 1/3 to 2/3 as long as the sessile spikelets; trigonous and subclavate, and with long hairs on one side. The *spikelets* are linear-oblong, glabrous or villous, 1/8 to 1/4 inch long, sessile and stalked spikelets close together; the pedicel of the stalked spikelet is thick about 1/3 or less than the length of the sessile spikelet, ciliate on one side, confluent with the thick callus of the sessile spikelet, which is sparsely bristly. The *sessile spikelet* consists of four glumes and is awned. The *first glume* is concave, pale yellow, shining and cartilaginous to about 2/3 its length from the base, and the upper third is membranous, dimidiately ovate; at the back in the cartilaginous portion, there are three to six deep convex smooth ridges running across the glume; the membranous tip is thin and with anastomosing green veins; the margins of this glume are thick, narrowly incurved, ciliolate, and with a narrow wing on the outer margin. The *second glume* is oblong-lanceolate or lanceolate, coriaceous, acuminate, scaberulous, keeled and laterally compressed and on the keel just below the tip there is a narrow ciliate wing. [Pg 154] The *third glume* is ovate-lanceolate, hyaline, acuminate 1- to 3-veined, male or empty, with a narrow hyaline palea. The *fourth glume* is shorter than the third, deeply cleft into two lanceolate acute lobes, 3-veined at the base; awn up to about 2/3 inch long; *palea* is linear lanceolate. Stamens are three and *lodicules* are small and cuneate.

The pedicelled spikelet is very variable. It is shorter than the sessile, with obscure transverse ridges and may consist of four glumes, but without an awn to the fourth glume; sometimes this spikelet is reduced to a single glume.

The grain is broadly oblong, brownish and compressed.

Distribution. — Throughout India and Ceylon.

[Pg 155]

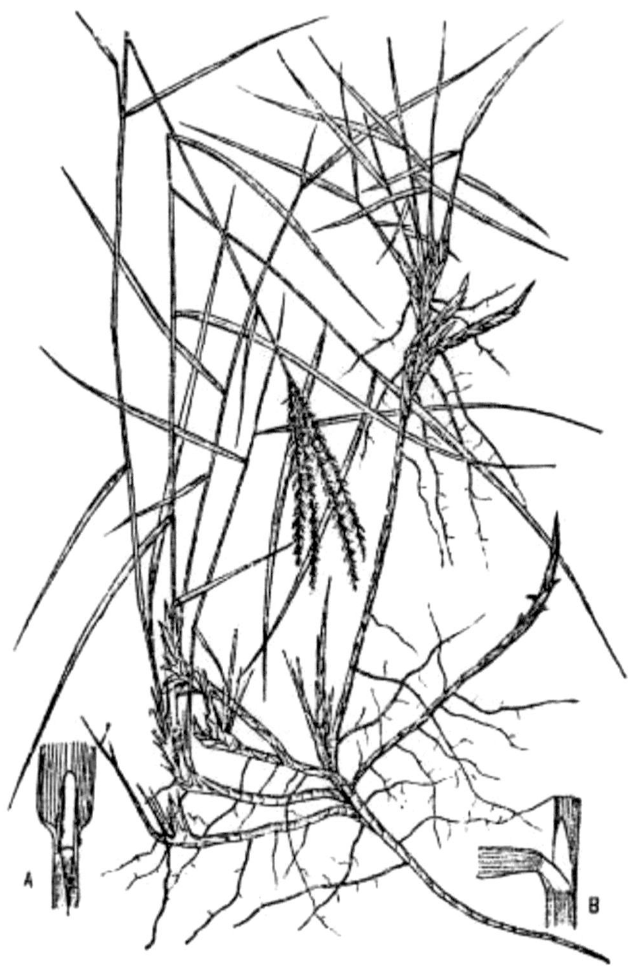

Fig. 133. — Ischæmum pilosum.
A and B. Ligules.

[Pg 156]

Ischæmum pilosum, *Hack.*

It is a tall, robust, perennial grass with rhizomes producing numerous creeping stolons densely covered with scaly-sheaths. The aerial stems are erect, freely branching at the base, slender, 2 to 3 feet long, glabrous.

The *leaf-sheath* is glabrous. The *ligule* is a distinct glabrous membrane, 1/8 inch long, rounded. *Nodes* are glabrous.

The *leaf-blade* is linear, finely acuminate, glabrous but bearded at the base, 6 to 12 inches long and 1/8 to 1/3 inch broad.

The *inflorescence* consists of two to six softly hairy spikes which are yellow or brown 1 to 4 inches long. Joints and pedicels are slender, sparsely ciliate.

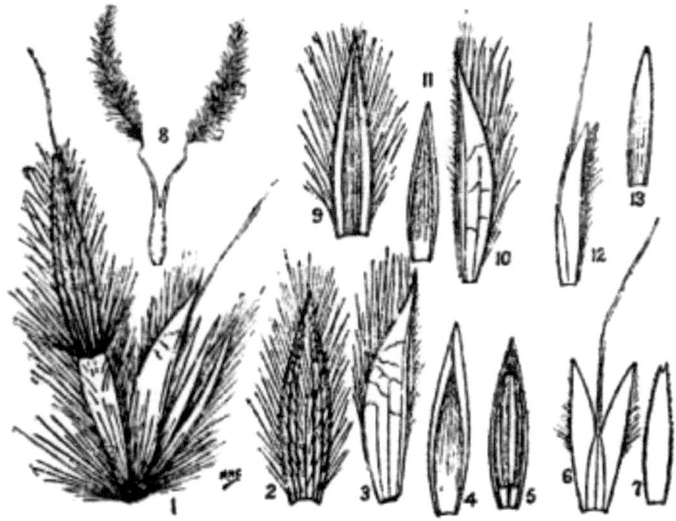

Fig. 134.—Ischæmum pilosum.
1. A sessile and a pedicelled spikelet; 2, 3, 4 and 6. the first, second, third and the fourth glume, respectively, of the sessile spikelets; 5. palea of the third glume, 7. palea of the fourth glume; 8. ovary; 9, 10, 11 and 12. first, second, third and the fourth glume, respectively, of the pedicelled spikelets; 13. palea of the fourth glume.

The *sessile spikelets* are narrowly lanceolate, 3/4 inch long, with long hairs at the base. The *first glume* is dorsally hairy, or glabrous, narrowed from the middle upwards, chartaceous, with incurved margins and six or seven anastomosing nerves. The *second glume* is longer than the first, laterally compressed, ovate-lanceolate, acuminate, chartaceous, glabrous but often with long hairs on the keel towards the upper half, 5-nerved, the lateral nerves anastomosing. The *third glume* is a little shorter than the second, linear-oblong or lanceolate, paleate; *palea* is membranous, nerveless, and encloses three stamens. The *fourth glume* is equal to the third glume in length, membranous, hyaline and divided almost to the middle into two acute lobes with an awn 1/4 to 3/8 inch long, paleate; *palea* is lanceolate, nerveless and encloses three stamens and the ovary and sometimes only the ovary. The *pedicelled* [Pg 157] *spikelets* are shorter than the sessile but with a shorter awn. The *glumes* are similar to those of the sessile spikelet; sometimes these spikelets are imperfect or even reduced to a single glume.

This grass grows well in black cotton soils and sometimes it gets very well established and then it is very difficult to eradicate it. Cattle seem to like this grass.

Distribution. — In black cotton soils all over the presidency, but most abundant in the Ceded districts.

[Pg 158]

Fig. 135. — Ischæmum ciliare.

[Pg 159]

Ischæmum ciliare, *Retz.*

It is a tufted perennial grass, erect or creeping. Stems are erect or ascending, sometimes decumbent at base, and rooting at the nodes, stout or slender, 6 inches to 2 feet long.

The *leaf-sheath* is compressed, loose, glabrous or hairy. The *ligule* is a short, ciliate membrane. *Nodes* are glabrous or hairy.

The *leaf-blade* is flat, linear-lanceolate, acuminate, narrowed towards the acute or rounded base, glabrous or hairy, 2 to 6 inches long and 1/6 to 1/2 inch wide.

The *inflorescence* consists of two spikes, 1-1/2 to 2 inches long; joints and pedicels of the pedicelled spikelets equal, hairy at the back and at the angles.

The *sessile spikelets* are 1/8 to 1/5 inch long, oblong, bearded at the base. The *first glume* is coriaceous, convex, polished, smooth or pitted, hairy below, flat and veined above the middle, with broad or narrow ciliate equal wings and with margins narrowly inflexed above and broadly so below. The *second glume* is coriaceous, equal to or longer than the first, lanceolate, acuminate, or shortly awned, 3- to 5-nerved, keel narrowly winged towards the apex, dorsally ciliate or not. The *third glume* is ovate-lanceolate, acuminate, ciliate towards the apex, 1- to 3-nerved, paleate; the *palea* has a coriaceous lanceolate centre, with broad hyaline ciliate wings and encloses three stamens. The *fourth glume* is hyaline, deeply lobed into two oblong obtuse glabrous or ciliate lobes, with an awn twice as long as the spikelet in the cleft, and paleate; *palea* is lanceolate, acuminate, 2-nerved. *Styles* and *stigmas* are short.

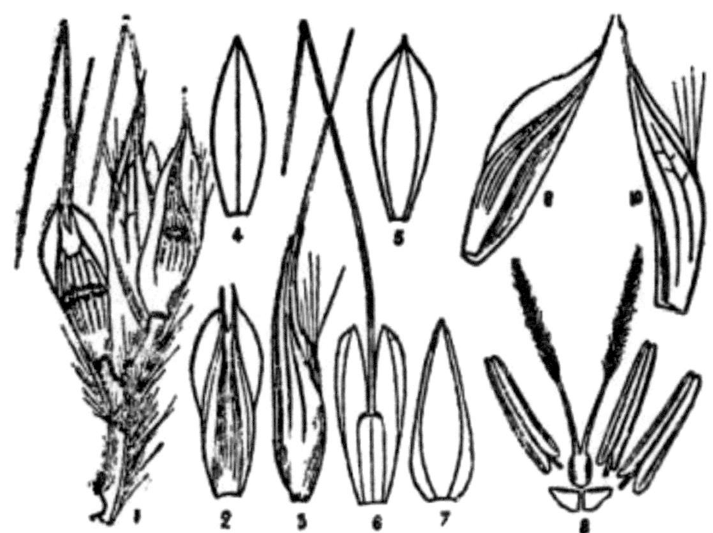

Fig. 136. — Ischæmum ciliare.
1. Spikelets; 2, 3, 4 and 6. the first, second, third and the fourth glume, respectively, of the sessile spikelet; 5 and 7. palea of the third and the fourth glumes, respectively; 8. lodicules, stamens and the ovary; 9 and 10. the first and the second glumes of the pedicelled spikelet.

The *pedicelled spikelets* resemble the sessile ones in the structure of their glumes and palea.

This grass is very variable in its habit and in the structure of its spikelets. It grows mostly in wet situations, such as the bunds of paddy fields and tanks. Cattle eat the grass eagerly.

Distribution. — All over India and Ceylon.

[Pg 160] Ischæmum laxum, *L.*

This is a perennial grass with numerous stiff, thick and wiry roots.

Stems are erect, slender, rising in tufts from a short root-stock, glabrous, leafy towards the base, varying in length from 2 to 3 feet.

The *leaf-sheaths* are shorter than the internodes usually glabrous, but occasionally with scattered hairs. At the mouth tufts of hairs are present or not. The *ligule* is a ridge of silky hairs. The *nodes* are glabrous.

The *leaf-blades* are erect, flat, slightly glaucous, linear, narrowed to long capillary tips, 5 to 12 inches long and 1/10 to 1/6 inch broad, with prominent nerves and scabrid margins.

The *inflorescence* is a solitary spike, 2 to 5 inches long, erect and fragile; the joints and pedicels are compressed, somewhat 2-angled, ciliate with long hairs, and about half as long as the spikelets.

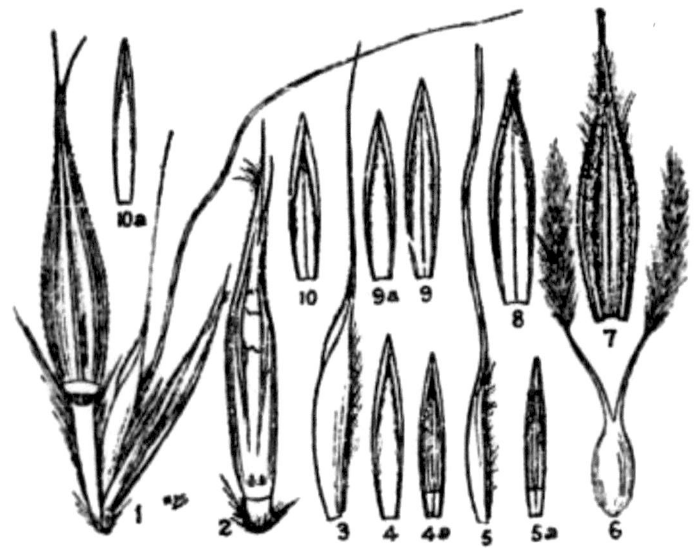

Fig. 137. — Ischæmum laxum.
1. A sessile and a pedicelled spikelet; 2. first glume of a sessile spikelet; 3. second glume of a sessile spikelet; 4 and 5. third and fourth glumes of sessile spikelets; 4a and 5a. are palea of third and fourth glumes; 6. ovary; 7, 8, 9 and 10. glumes of pedicelled spikelets; 9a and 10a. palea of third and fourth glumes.

The *spikelets* are in pairs, one-sessile and one-pedicelled. The *sessile spikelets* are pale-green, linear-oblong, acuminate with a shortly bearded callus, 1/4 to 3/8 of an inch long. There are four glumes in a spikelet. The *first glume* is chartaceous, oblong-lanceolate, acuminate, 2-toothed with the teeth ending in two short awns, densely ciliated at the apex on one side, conspicuously 6- (rarely) 7-nerved, the two lateral being very strong and running into the apical teeth and the intermediate four nerves being shorter and not running up to the apex, and on the dorsal surface there is a depression, where it is membranous and the nerves on its sides sometimes anastomosing at the upper third of the glume. The *second* [Pg 161] *glume* is shorter than the first, chartaceous to a certain extent, ovate-lanceolate, acuminate, concave, terminating in a fine scabrid awn, 1/2 inch long, with margins ciliate from above the middle to the apex, and with a narrow ciliated wing on the keel at the apex running up to the base of the awn, 3-nerved. The *third glume* is lanceolate, acuminate, hyaline, nerveless, ciliate, with a linear obtuse *palea* enclosing three stamens and two *lodicules*. The *fourth glume* is hyaline, membranous, deeply split at the apex into two prominent lobes and with an awn in the depression 1/2 inch long, the *palea* is linear oblong and contains either the ovary alone or both the *stamens* and the *ovary*.

The *pedicelled spikelets* are also as long as the sessile, more conspicuous than the sessile and consist of four glumes, but are not awned. The *first glume* is lanceolate or oblong-lanceolate, chartaceous, with seven strong nerves, very prominent at the back and the mid nerve being most conspicuous, with scabrid keels and closely finely ciliated and folded margins, finely biaristate at the apex. The *second glume* is lanceolate, finely acuminate, sub-chartaceous, with the margins ciliate from about two-third its length from the apex, 3-nerved, the mid nerve alone being prominent. The *third glume* is hyaline, nerveless, lanceolate, ciliate in the margin, paleate with 3 stamens or empty. The *fourth glume* is shorter than the third, hyaline, narrow lanceolate, not awned, ciliate or not at the margin, paleate and with three stamens and two *lodicules*.

This grass produces a large amount of leaves in good soils and it is liked very much by cattle. It is capable of standing a long spell of dry weather, and is valuable in this respect because it can be depended upon when other grasses fail. It is worth conserving with other grasses. It grows both in rich and poor soils, in open places and also in thickets.

Distribution. — Throughout India and Ceylon.

[Pg 162]

21. Eremochloa, *Buse.*

These are tufted perennial grasses with rigid equitant leaves at the base. The inflorescence consists of a solitary, glabrous, and compressed spike, with a somewhat fragile rachis; the joints are compressed, hollow and clavate. The spikelets are solitary, usually 2-flowered (rarely 1-flowered), secund, closely imbricating, sessile with a short, pedicelled, reduced upper spikelet, and deciduous with the joint. There are four glumes. The first glume is oblong or ovate, flat, smooth, coriaceous, pectinately margined with upcurved spines. The second glume is oblong-lanceolate, acute and 3-nerved. The third glume is hyaline, obtuse, paleate and male. The fourth glume is smaller, hyaline, oblong, obtuse, 1-nerved, paleate, bisexual or female. Lodicules are truncate and slightly oblique. Stamens are three with long anthers. Styles are two with feathery stigmas. The grain is oval, plano-convex.

[Pg 163]

Eremochloa muricata, *Hack.*

This is a perennial tufted grass with a woody creeping root-stock. Stems are erect or ascending, slender, strongly compressed, lower parts completely covered by rigid equitant leaves, 6 to 18 inches long or more.

The *leaf-sheath* is broad, flat, much compressed, glabrous and keeled. The *ligule* is a short membrane. *Nodes* are glabrous.

The *leaf-blade* is linear, glabrous on both sides, 2 to 6 inches long and 3/16 to 1/4 inch broad, with a rounded tip and two unequal lobes.

The *spike* is solitary, up to 6 inches (or more) in length, joints of the rachis 1/3 to 1/2 the length of the spikelets. *Spikelets* are solitary, sessile, compressed, secund. The *sessile spikelets* are 3/16 to 1/6 inch, and consist of four spikelets. The *first glume* is oblong-lanceolate, dorsally slightly convex, smooth, coriaceous, 7- to 9-nerved, and with pectinate margins consisting of long, spreading, upcurved

230

spines and at the top with subquadrate wings on each side reaching beyond the acute tip. The *second glume* is chartaceous, oblong-lanceolate, acuminate, usually 5-nerved (and occasionally 3-nerved), the mid-nerve keeled with a narrow wing from below the middle to the base and with hyaline margins. The *third glume* is oblong-obovate, hyaline, thin, paleate with three yellow *anthers* and two oblong-cuneate *lodicules*; *palea* is narrow, oblong, obtuse. The *fourth glume* is thin, hyaline, oblong-lanceolate, obtuse, paleate, and bisexual; *palea* lanceolate, narrow, two-toothed at the apex, with deep purple *anthers* and *stigmas* of the same colour. The *lodicules* are obliquely truncate. *Ovary* has a reddish spot between the style branches and just at the apex in the fresh state in the bud and in the open flower.

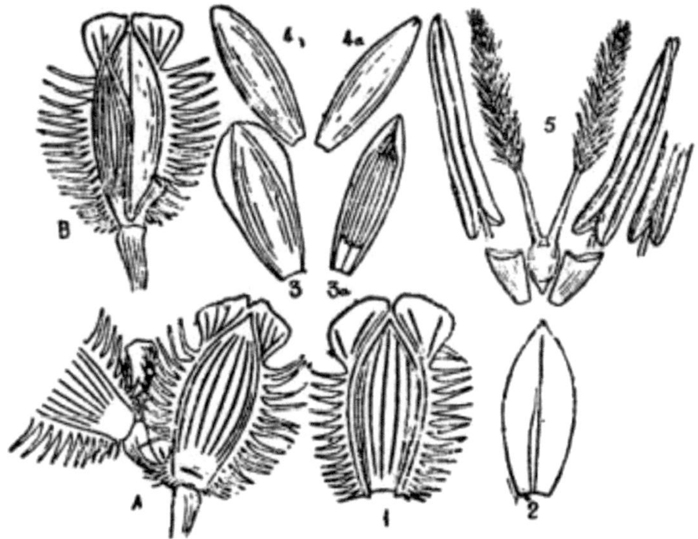

Fig. 138. — Eremochloa muricata.
A. Sessile spikelet; B. sessile and a pedicelled spikelet; 1, 2, 3 and 4. the first, second, third and the fourth glume of the sessile spikelet; 3a and 4a. palea of the third and fourth glumes of the sessile spikelet; 5. ovary, anthers and lodicules.

The *pedicelled spikelet* is reduced to an inflated body, as long as the sessile spikelet. It is pointed towards both ends, green with anastomosing veins on the outside and membranous, white and nerveless on the other side. The part of the pedicelled spikelet corresponding to the spikelet looks as if the margins of the first and second glumes are confluent all round.

Distribution. — South India and Ceylon.

[Pg 164]

22. Apocopis, *Nees.*

These are annual or perennial grasses with slender stems. The spikes are compressed, 2- to 3-nate, or solitary at the ends of slender branches, with a rachis not jointed; joints are short, slender and villous. Spikelets are closely imbricating in two series, sessile, solitary, the upper reduced to a small pedicel 1- to 2-flowered, the lowest few on the spike, imperfect, male or neuter. There are four glumes. The first glume is large, broadly obovate or obcordate, cuneate, villous with brown hairs, 7- to 9-nerved. The second glume is as long as the first, but narrower, thinner, oblong to ovate, spikelet truncate and 3-nerved. The third glume is hyaline, narrow, paleate, male or empty. The fourth glume is hyaline, linear, entire or 2-fid, awned, bisexual with a very short palea. Lodicules are absent. Stamens are two or three with linear anthers. Styles are short and stigmas slender and exserted. The grain is small, oblong and narrow.

[Pg 165]

Fig. 139. — Apocopis Wightii.

[Pg 166]

Apocopis Wightii, *Nees ex Steud.*

This is a low and densely tufted or tall erect annual grass. Stems are leafy, branching freely, 3 to 8 inches long.

The *leaf-sheath* is loose, usually hairy, rarely also glabrous and hairy at the mouth. The *ligule* is a small lacerate membrane.

The *leaf-blade* is linear-lanceolate, acuminate, hairy on both sides and with tubercle-based hairs, rarely glabrous, 3/4 to 3 inches by 1/12 to 1/8 inch.

The *inflorescence* consists of two racemes, closely appressed together on a very slender peduncle; the joints are shorter than the spikelets and with long brown hairs.

The *spikelets* are oblong, 1/8 to 1/5 inch long, the callus is short, hairy with long brown hairs. The *first glume* is cuneately obovate or obcordate, yellowish with red brown tips or dark brown with yellow tips, chartaceous below, membranous, hyaline and ciliate at the truncate, emarginate or retuse apex, 7- to 9-nerved, the nerves abruptly ceasing towards the apex. The *second glume* is as long as the first, broadly oblong, sides sharply folded inwards, 3-nerved, rarely nerveless, with long hairs at the back towards the base and with short cilia at the apex. The *third glume* is as long as the first, hyaline, thin, linear-oblong, nerveless, ciliate at the apex, paleate, usually with two stamens or empty; *palea* as long as the glume, hyaline and nerveless. The *fourth glume* is slightly longer than the other glumes or equal, very narrowly oblong or linear, membranous, awned and paleate; *awn* is 2 to 6 times the length of the glume, 7/8 to 1-1/4 inch long; *palea* is hyaline, thin, nerveless, convolute, broadly oblong to almost quadrate oblong, apex with very short cilia. Grain is minute and oblong.

Fig. 140. — Apocopis Wightii.
1. Spike; 2. a spikelet; 3, 4, 5 and 7. the first, second, third and the fourth glume, respectively; 6 and 8. palea of the third and the fourth glume; 9. ovary.

This grass varies very much in its spikelets. In one form they are smaller and hairy and in the other they are larger and glabrous except for a few stray hairs here and there. The former one is more widely distributed and the latter seems to be confined to certain localities in the south of the Presidency.

Distribution.—Throughout the Deccan Peninsula, Behar, Central India, Burma and Ceylon.

[Pg 167]

23. Lophopogon, *Hack.*

These are small densely tufted perennial grasses, with very narrow leaves. The spikes are very short at the ends of very fine branches, solitary, binate or fascicled, with very fragile rachis; joints are very short, slender with cupular tips. The spikelets are binate one sessile and the other shortly pedicelled, with the callus villous. There are four glumes. The first glume (of both the sessile and the pedicelled spikelets) is oblong, truncate, irregularly 3- to 4-toothed, 5- to 7-nerved and dorsally convex. The second glume is narrow lanceolate, longer than the first, 3- to 5-nerved, hispidly villous dorsally below the middle and on the sides, aristate or awned. The third glume is oblong lanceolate, hyaline, acute or aristate, 1-nerved, male or neuter, with a linear palea. The fourth glume is hyaline, as long as the third, entire or 2-fid and awned in the pedicelled and not awned in sessile spikelets, paleate with female or bisexual flowers. Lodicules are not present. Stamens are two. Stigmas are long.

[Pg 168]

Lophopogon tridentatus, *Hack.*

This is a small annual grass with slender, tufted, erect stems varying in height from 4 to 12 inches.

Leaf-sheaths are glabrous or with scattered hairs. The *ligule* is a fringe of close-set long hairs. *Nodes* are covered with long hairs below, but nodes nearer the inflorescence are glabrous.

Leaf-blades are very finely linear, acuminate, rigid, erect, glabrous below, with long hairs on the upper surface to about quarter the length of the blade and densely hairy near the mouth, and varying in length from 2 to 6 inches.

The *inflorescence* consists of usually two closely appressed spikes, though appearing as one, 1/2 to 3/4 inch long, pilose with ferrugineous hairs; the peduncle is capillary and enclosed by the upper leaf-sheath.

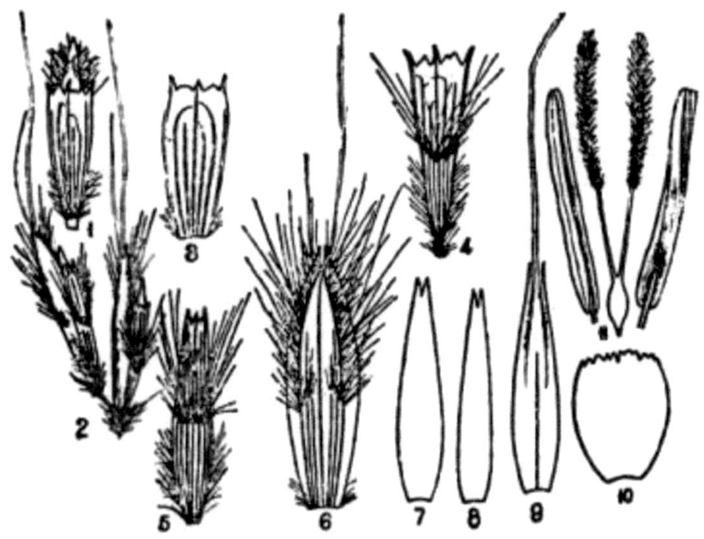

Fig. 141.—Lophopogon tridentatus.
1. Awnless lower spikelet; 2. a lower sessile and an upper pedicelled spikelet; 3. the first glume of an awnless lower spikelet; 4. the first glume of a lower spikelet that is awned; 5. the first, glume of the upper or pedicelled spikelets; 6, 7 and 9. the second, third and the fourth glume, respectively, of the upper pedicelled spikelets; 8 and 10. palea of the third and the fourth glumes; 11. ovary and anthers.

The *spikelets* are densely imbricate, binate at each joint, the upper being shortly pedicelled and the lower sessile or subsessile. The *lower spikelets* are 1/5 inch long with a tuft of brownish hairs at the tip of the callus. The *lower spikelets* at the very base of the inflorescence are awnless and contain only two male flowers, whereas those above in the inflorescence are awned and contain one male flower and one hermaphrodite or female flower.

There are four *glumes* in the spikelet. The *first glume* in the awnless spikelets is coriaceous, oblong, cuneate, very sparsely hairy or glabrous, shorter than the second glume, 7-nerved, 5-toothed at the

236

apex, two teeth being broader and shorter and three sharper and longer. The *second glume* is longer than the first, 1/5 inch long, sub-chartaceous, lanceolate, 3-nerved, 2-fid at the tip and awned or aristate, margin hyaline and with long [Pg 169] brownish hairs on the marginal nerves. The *third glume* is hyaline, a little shorter than the second, lanceolate-linear, tip bifid or irregularly toothed, paleate with two stamens or rarely empty; the *palea* is linear, about as long as the glume, tip irregularly toothed. The *fourth glume* is hyaline, as long as the third glume, 2-fid at the tip, awnless with a very minute arista in the cleft or not, paleate with two stamens; *palea* narrow and hyaline. The *first glume* of the lower spikelets above is somewhat narrower, 5- or 3-toothed with long hairs at the margins and with tufts of hairs at the back about the middle. The pedicelled or upper spikelets also have four *glumes* and bear one male flower and one bisexual flower. The *first glume* is shorter than the second glume, narrow, oblong, cuneate, 3-toothed with marginal hairs and tufts of hairs at about the middle at the back, 7-nerved all nerves running straight. The *second glume* is longer than the first, 1/5 inch long, sub-chartaceous, lanceolate, 2-fid at the tip, awned with hyaline margins, 3- to 7-nerved, marginal nerves with long brown hairs, and also with two tufts of hairs at about the middle or without it. The *third glume* is hyaline, nerveless, linear-lanceolate, shorter than the second glume, tip irregularly toothed or unequally bifid, paleate with two stamens; *palea* is linear about as long as the glume. The *fourth glume* is hyaline, about 1/6 inch long, lanceolate, 2-fid at the tip, awned in the cleft, lobes are hairy; *awn* is 3/4 inch long, paleate, usually bisexual, rarely female; *palea* is two-thirds of the glume in height, broadly ovate or quadrate, lobulate at the apex. *Styles* are very long, purple, *anthers* long, yellow. Grain narrow ellipsoidal or cylindric as long as the palea.

This grass is found in Chingleput, Nellore and Chittoor districts in open waste places in loamy soils.

Distribution. — The Konkan, Kanara and Central Provinces.

[Pg 170]

24. Apluda, *L.*

These are tall leafy slender perennial grasses, with branching stems erect or geniculately ascending from a creeping or decumbent base. The inflorescence is a leafy panicle of many small spikes enclosed in spathiform bracts. Spikes are of one linear joint gibbously bulbous at the base, and jointed on the peduncle at the base of the spathe by a minute curved pedicel. Spikelets are three, a sessile, 2-flowered bisexual one in front, and two pedicelled ones behind, one of which is imperfect and reduced to a glume and the other perfect male or rarely bisexual. The two pedicels are flat, prolonged from one side of the rounded rachis, oblong linear, truncate with a few long hairs along the margin. Sessile spikelets have four glumes. The first glume is chartaceous, linear oblong, many-nerved, shortly bifid at the apex, longer than the other glumes. The second glume is thinner, dorsally gibbous, keeled, 5- to 9-nerved, beaked and minutely bifid. The third glume is hyaline, oblong, acute, 3-nerved, paleate and male. The fourth glume is hyaline, deeply bifid, awned in the sinus, bisexual with a minute palea. The pedicelled spikelet has also four glumes. The first and the second glumes are nearly equal, rather chartaceous. linear-oblong, acute or acuminate, many-nerved. The third glume is hyaline, oblong-lanceolate, 3-nerved, paleate and male. The fourth glume is hyaline, bifid, paleate, 1-nerved, female or bisexual. Lodicules are two. Stamens are three. Grain is oblong.

[Pg 171]

Apluda varia, *Hack.*

This is a tall leafy perennial grass with wiry roots. Stems are densely tufted, branched, geniculately ascending, erect or the branches scandent, solid, smooth and polished, 1 to 7 feet.

The *leaf-sheath* is glabrous or slightly hairy, the upper ones being shorter and dilated into spathes with subulate tips. The *ligule* is a short stiff slightly lacerate membrane.

The *leaf-blade* is linear-lanceolate, finely acuminate, base narrowed into a petiole, scaberulous on both the surfaces.

The *inflorescence* consists of simple spikes, each in a spathiform bract, and forming clusters terminating the stem and the branches. The *spikes* have their bases rounded and swollen and each spike consists of a sessile bisexual spikelet and two flat linear, truncate, parallel pedicels, one terminated by a spikelet, and the other by a solitary minute glume. Spathes are 1/8 to 1/3 inch long, sessile or pedicellate, green, cymbiform, with subulate tips.

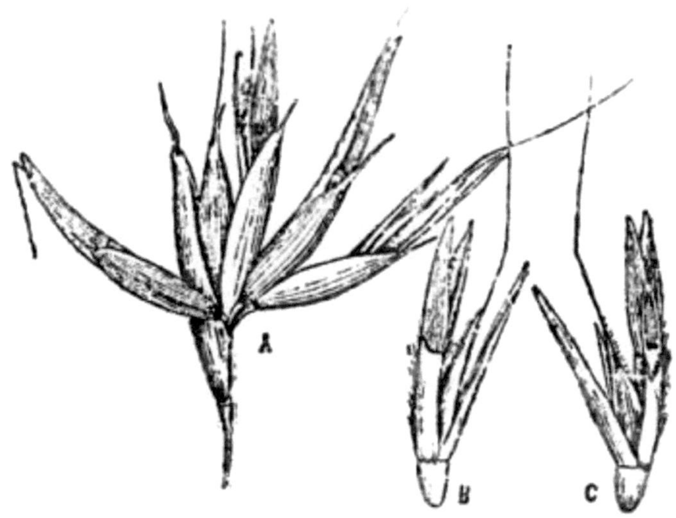

Fig. 142.—Apluda varia.
A. A cluster of spikes containing five spikes with their spathes; B and C. spikes.

The *sessile* as well as the *pedicelled spikelets* have four *glumes*. The *sessile spikelets* are 1/8 to 1/5 inch long. The *first glume* is spreading or erect, chartaceous, many-nerved, two-toothed at the apex and with narrow hyaline margins from about the middle to the apex. The *second glume* is compressed, dorsally gibbous, keeled, 7-nerved. The *third glume* is hyaline, oblong-lanceolate, 3-nerved, paleate with three stamens; *palea* is narrow. The *fourth glume* is shorter than the third, deeply 2-fid and awned in the cleft, bisexual or female, 3- to 5-nerved below the cleft, the lateral nerves arching and meeting the mid-nerve just at the cleft, with a small ovate palea. There are two *lodicules*. The *pedicelled spikelets* are dorsally compressed. The *first glume* is lanceolate, oblong, subacute, many-nerved, coriaceous and glabrous. The *second glume* is as long as the first, many-nerved, lanceolate-oblong, coriaceous and glabrous. The *third glume* is hyaline, shorter than the second, 3-nerved, paleate and with three stamens. The *fourth glume* is shorter than or equal to the third, hyaline, 1-nerved rarely with two short lateral nerves, female or imperfect. *Lodicules* are two.

[Pg 172]

A very common grass occurring in the plains and lower hills, all over the Presidency and grows well in all kinds of soil.

Distribution. — All over India.

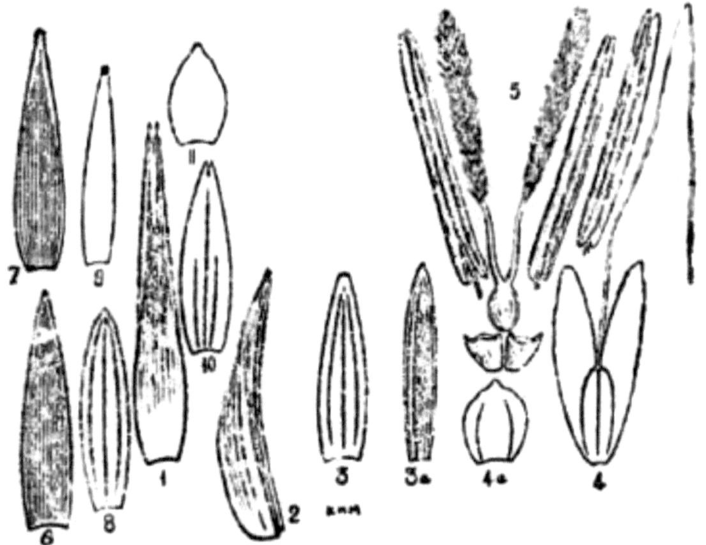

Fig. 143. — Apluda varia.
1, 2, 3 and 4. The first, second, third and the fourth glume, respectively, of the sessile spikelet; 3a and 4a. are the palea of the third and the fourth glume, respectively; 5. stamens, ovary and lodicules; 6, 7, 8 and 10. the first, second, third and the fourth glume, respectively, of the pedicelled spikelet; 9 and 11. palea of the third and the fourth glumes.

240

[Pg 173]

25. Rottboellia, *Linn. f.*

These are tall, annual or perennial grasses, with leafy stems and narrow leaves. The spikes are few or many, solitary or panicled, with a jointed usually fragile rachis; the joints are rounded or compressed, hollowed on one side and excavated at the tip. The spikelets are usually binate, one-sessile closing or sunk in the cavity of the joint and the other pedicelled, smaller than the sessile or rudimentary with the pedicel usually adnate to the joints and equal to or shorter than it. The sessile spikelets are bisexual, 1- to 2-flowered, equal to or shorter than the joint and four-glumed. The first glume is coriaceous dorsally flattened, obtuse, margins narrowly incurved. The second glume is thinner than the first, broadly ovate, acute and gibbously convex. The third glume is hyaline, ovate, acute, male or neuter, with a membranous palea. The fourth glume is hyaline, bisexual, broadly ovate, acute with a hyaline, ovate-lanceolate palea. There are three stamens with linear anthers. There are two cuneate lodicules. Styles are two with laterally exserted stigmas. The grain is broadly oblong. The pedicelled spikelets are smaller than the sessile, male or neuter, with four glumes. The first glume is herbaceous, many-nerved, ovate-acute, minutely bifid at the apex. The second, third and the fourth are more or less similar to those of the sessile spikelet.

KEY TO THE SPECIES.

- Spike solitary, the first glume of the sessile spikelet broadly winged. R. Myurus.
- Spikes fascicled, the first glume of the sessile spikelet narrowly winged. R. exaltata.

[Pg 174]

Rottboellia Myurus, *Benth.*

This is a tufted perennial with creeping stems which branch freely into ascending compressed branches, 10 inches to 2 feet high.

The *leaf-sheath* is quite glabrous and compressed. The *ligule* is a short ciliate membrane. *Nodes* are glabrous.

The leaf-blade is flat, linear, acute, glabrous, 2 to 6 inches long.

The *inflorescence* consists of a solitary terminal or axillary *raceme* 1 to 2 inches long; joints are shorter than the spikelets, excavate on one side and with a pore which is hidden by the sessile spikelet. The *sessile spikelet* consists of *four glumes*. The *first glume* is somewhat fiddle-shaped, dilated above the middle into an orbicular wing, and towards the base into two auricles joined by a transverse ridge, scaberulous, 5-nerved. The *second glume* is somewhat membranous, ovate, acute and 3-nerved. The *third glume* is hyaline, thin, oblong, obtuse and nerveless. The *fourth glume* is lanceolate, nerveless and without a palea, bisexual. There are two cuneate *lodicules*. The *pedicelled spikelets* also have four glumes and the pedicels usually free, but also sometimes adnate. The *first glume* is oblong, obtuse, winged on one side only, 5-nerved. The *second glume* is boat-shaped, chartaceous, 3-nerved crested with a semi-circular wing at the apex. The *third glume* is hyaline, broadly oblong, obtuse, 3-nerved with a lanceolate hyaline palea. The *fourth glume* is oblong, obtuse, male.

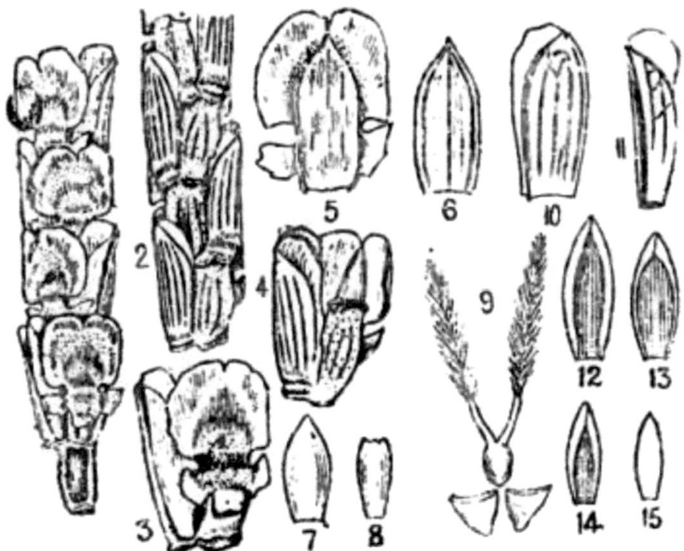

Fig. 144.—Rottboellia Myurus.

1. A portion of the raceme showing front view; 2. a portion of the raceme showing the back view; 3. a sessile and a pedicelled spikelet showing the front side; 4. the same showing the back side; 5, 6, 7 and 8. the first, second, third and the fourth glume of the sessile spikelet, respectively; 9 ovary and lodicules; 10, 11, 12 and 14. the first, second, third and the fourth glume, respectively, of the pedicelled spikelet; 13 and 15. palea of the third and fourth glumes of the sessile spikelet.

This is very common in dry somewhat sandy places in the East Coast districts.

Distribution.—Common in Deccan peninsula.

[Pg 175]

Rottboellia exaltata, *L.f.*

This grass is usually annual and rarely perennial. Stems are stout, erect, hispid, branching from the base, varying in height from 3 to 10 feet.

The *leaf-sheaths* are loose, hispid with tubercle-based hairs, or glabrous, with mouth contracted. The *ligule* is short and ciliate.

The *leaf-blade* is linear-lanceolate, setaceously-acuminate with a stout midrib prominent beneath, hispid or scabrid above, smooth or sometimes scaberulous and glaucous beneath, spinulosely scabrid at the margin, 5 to 24 inches by 1/4 to 1 inch.

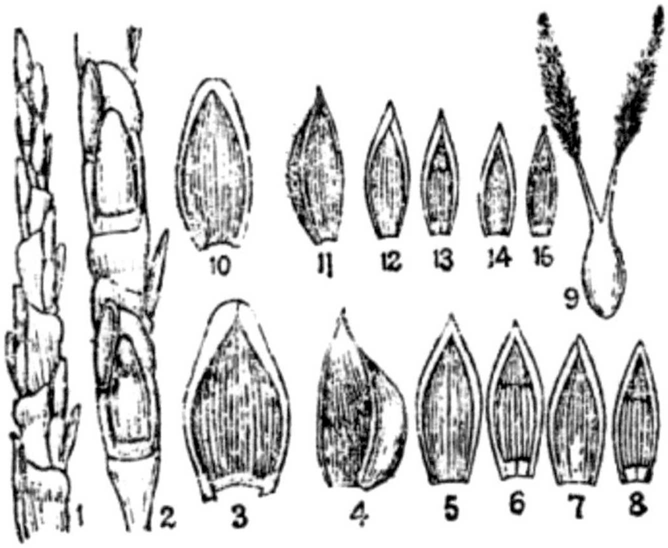

Fig. 145. — Rottboellia exaltata.
1 and 2. A portion of the spike, back and front view; 3, 4, 5 and 7. the first, second, third and the fourth glume, respectively, of the sessile spikelet; 6 and 8. palea of the third and the fourth glumes of the sessile spikelet; 9. ovary; 10, 11, 12 and 14. the first, second, third and the fourth glume of the pedicelled spikelet; 13 and 15. palea of the third and the fourth glume of the pedicelled spikelet.

The *racemes* are stout, cylindrical below and very narrow and with imperfect spikelets above, joints are smooth and rounded dorsally. The *sessile spikelets* are as long as the joint or slightly shorter and has four glumes. The *first glume* is ovate-oblong, thickly coriaceous, smooth at the back with a truncate base and a transverse ridge at the base inside, many-nerved, with very narrow inflexed margins and very narrow wings at the top, the apex is obtuse or emarginate. The *second glume* is equal to the first glume in height, chartaceous, gibbously convex, broadly ovate, acute, 9- to 11-nerved, and with a short wing to the keel at the apex. The *third glume* is oblong or elliptic-oblong, rigid with a hyaline centre and coriaceous at the sides, 3-nerved, paleate and with three stamens; *palea* is as long as the glume, coriaceous with inflexed hyaline margins. *Lodicules* are cuneate, with toothed edge. The *fourth glume* is a little shorter than the third, ovate from a broad base, hyaline and acute, 1-nerved, paleate and usually with an ovary and two *lodicules*: *palea* is hyaline, as long as the glume. [Pg 176] but narrower, nerveless. *Lodicules* are quadrate; grain somewhat large oblong and compressed. The *pedicelled spikelets* are usually imperfect.

This grass occurs all over the Presidency in cultivated dry fields.

Distribution.—Throughout the lower hills and plains of India and in Australia and Africa.

[Pg 177]

26. Mnesithea, *Kunth.*

These are erect slender perennial grasses with narrow leaves. The spikes are solitary and slender, with a fragile, articulated rachis; the joints are terete, ribbed, all but a few upper with two equal and similar sessile spikelets, sunk in sub-opposite oblong cavities, separated by a hyaline septum, and with sometimes a minute glume representing a third spikelet (the pedicelled) on the upper margin of the joint. The sessile spikelets are one-flowered, nearly as long as the internode. There are four glumes in the spikelet. The first glume closing the mouth of the cavity in the joint is obliquely oblong, obtuse, smooth with narrowly incurved margins. The second and the third glumes are as long as the first, obtuse and hyaline. The third glume is empty, paleate or not. The fourth glume is rather small, oblong, obtuse, bisexual and palea shorter than the glume. The lodicules are not present. The stamens are three. Ovary is very small with stigmas not exserted. The grain is narrowly oblong compressed. The pedicelled spikelets are confined to the upper 1-flowered joints of the spike and their pedicels are confluent with the walls of the joints and their margins are marked by two ribs. The first glume is very minute and the other glumes are absent.

[Pg 178]

Mnesithea lævis, *Kunth.*

This is an erect slender perennial grass with smooth simple or branched stems varying in height from 2 to 4 feet.

The *leaf-sheath* is terete, tight, glabrous. The *ligule* is a short toothed membrane. *Nodes* are glabrous.

The *leaf-blade* is flat, linear from a narrow base, glabrous or base hairy; apices of upper leaves acuminate, and those of the lower obtuse, with finely serrate margins and a midrib prominent below, 6 to 12 inches long and 1/10 to 1/6 inch wide.

Racemes are short, exserted from the uppermost sheath, erect, 4 to 8 inches long; joints are 1/5 inch long, contracted in the middle, with two equal and similar spikelets, sunk in the opposite oblong cavities separated by a thin hyaline septum and sometimes with a minute glume of the third spikelet on the upper margin of the joint.

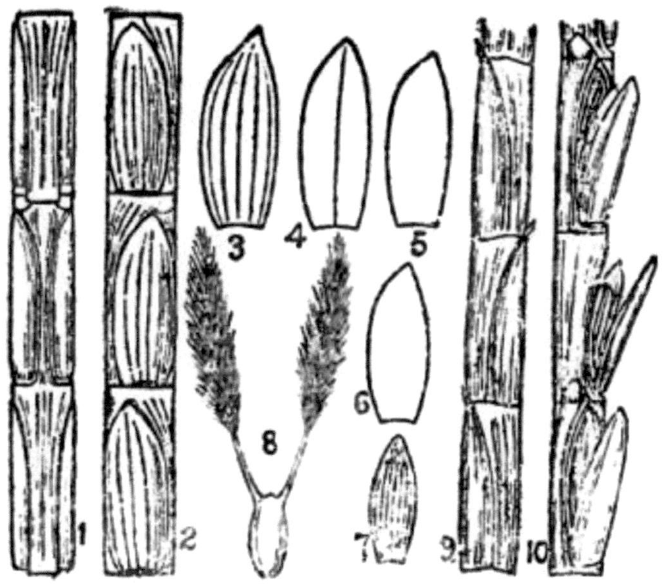

Fig. 146.—Mnesithea lævis.
1 and 2. Portions of a spike; 3, 4, 5 and 6. the first, second, third and the fourth glume, respectively; 7. palea of the fourth glume; 8. ovary; 9 and 10. a part of the spike at the terminal portion.

The *sessile spikelets* are 1-flowered, as long as the joint and varying in length from 1/7 to 1/5 inch and have four *glumes*. The *first glume* is obliquely oblong, coriaceous, smooth, obtuse, margins narrowly incurved, truncate and pitted at the base, 5- to 7-nerved. The *second glume* is as long as the first hyaline, oblong and obtuse. The *third glume* is like the second but thinner and slightly broader, paleate or not, empty. The *fourth glume* is rather smaller than the third, oblong, obtuse, bisexual and paleate; the *palea* is shorter than the glume. *Lodicules* are not present.

This grass is usually found in dry fields all over the presidency but it is nowhere abundant.

Distribution.—Throughout India and Ceylon.

[Pg 179]

27. Manisuris, *Sw.*

These are erect leafy much branched annual grasses. Leaves are amplexicaul and cordate at the base. The inflorescence consists of small, terete, axillary and terminal spikes with peduncles often confluent in a leafy spiciform panicle; the rachis is fragile with short broad joints, deeply excavate opposite the sessile spikelets and the tips with two pits. Spikelets are in dissimilar pairs, one globose, sessile and bisexual and the other ovate, pedicelled, neuter with the pedicels adnate to, or closely appressed to the joint of the rachis. The sessile spikelet has four glumes. The first glume is globose, hard, coarsely pitted, with an oblong ventral opening opposite the cavity in the joint of the rachis. The second glume is chartaceous, minute, oblong, 1-nerved immersed in the cavity of the first glume and closing the opening. The third and the fourth glumes are hyaline and minute. The lodicules are broadly cuneate. Anthers are minute. The styles are free and stigmas are short exserted from the opening in the first glume. Grain is sub-globose.

[Pg 180]

Manisuris granularis, *L.f.*

This is a freely branching annual with stems leafy to the top and varying in length from 1 to 2-1/2 feet.

The *leaf-sheath* is inflated, covered with scattered tubercle-based hairs. The *ligule* is a short membrane with ciliate margin. *Nodes* are with long hairs.

The *leaf-blade* is linear, cordate and amplexicaul at base, acute, flat, flaccid, with scattered tubercle-based hairs on both the surfaces, 4 to 10 inches by 1/4 to 1/2 inch.

The *spikes* are solitary, axillary and terminal and 1/4 to 1 inch, the peduncles of the spikes are often confluent in a leafy spathiform panicle; the rachis is fragile with short joints deeply excavate on one side.

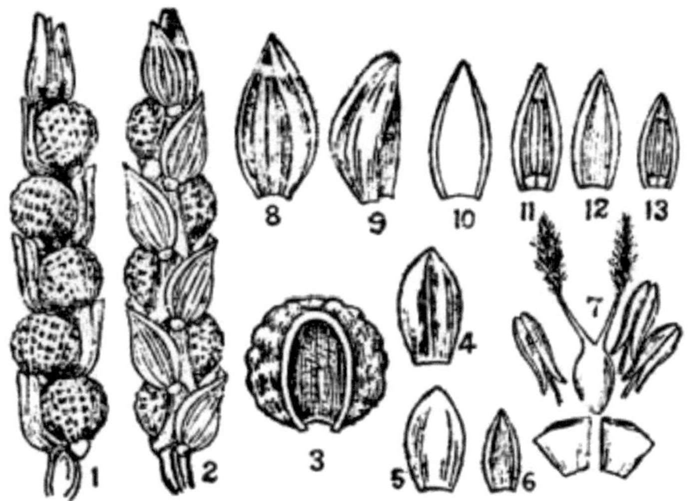

Fig. 147.—Manisuris granularis.
1 and 2. The front and back view of a bit of a spike; 3, 4, 5 and 6. the first, second, third and the fourth glume, respectively, of the sessile spikelet; 7. ovary anthers and lodicules; 8, 9, 10 and 12. the first, second, third and the fourth glume, respectively, of the pedicelled spikelet; 11 and 13. palea of the third and fourth glumes.

The *spikelets* are 1- to 2-flowered in dissimilar pairs, one globose, sessile and bisexual and the other ovate, pedicelled, neuter; the pedicel is adnate to the joint of the rachis.

The *sessile spikelet* has four *glumes*. The *first glume* is hard, globose, foveolate, with an oblong opening, faintly nerved. The *second glume* is chartaceous, immersed in the cavity of the joint, and filling the opening. The *third glume* is small hyaline and empty. The *fourth glume* is hyaline, small and paleate. The grain is sub-globose. *Lodicules* are broadly cuneate.

The *pedicelled spikelets* also have four *glumes*. The *first glume* is ovate, sub-chartaceous, winged on one side with a broad hyaline ciliate wing, 5- to 7-veined. The *second glume* is cymbiform, compressed laterally, with a dorsal hyaline ciliate wing to the keel, 5- to 7-veined. The *third glume* is hyaline, membranous, oblong, 2-nerved

249

and paleate or not, and with or without stamens. [Pg 181] The *fourth glume* is similar to the third, but slightly smaller, paleate and with three stamens.

This grass occurs in open loamy soils and in cultivated dry fields.

Distribution. — Throughout India and Ceylon and also in most of the tropical countries.

[Pg 182]

28. Andropogon, *L.*

The grasses of this genus are either perennial or annual and vary very much in habit. The inflorescence consists of solitary, binate, digitate, or panicled racemes. The rachis is usually jointed and fragile. Spikelets are binate, a sessile female or bisexual and a pedicelled male or neuter. The sessile spikelet is 1-flowered and has usually four glumes. The first glume is coriaceous or chartaceous, dorsally compressed, with incurved margins, usually 2-keeled. The second glume is as long as the first, thinner, with a median keel, laterally compressed, awned or not. The third glume is hyaline, empty, nerveless and without a palea. The fourth glume is hyaline, narrow or broad, 2-fid and awned, or reduced to an awn more or less dilated at the base, paleate or not. There are two lodicules and three stamens. Stigmas are feathery. Grain is free. The pedicelled spikelets are usually smaller than the sessile and have three or four glumes and are awnless.

KEY TO THE SPECIES.

- A. Sessile spikelets all similar.
    - B. Racemes of many spikelets.
        - C. Peduncle of racemes enclosed in spathiform leaf-sheaths.
            - D. Joints of rachis and pedicels of upper spikelets slender and tips obliquely truncate.
                - Racemes solitary, pedicelled spikelets similar to the sessile, glume 1 of

sessile spikelets pitted. 1. A. foveolatus.

- DD. Joints of rachis and pedicels of upper spikelets clavate or trumpet-shaped and tips cupular with toothed margins.
    - Racemes binate, pedicelled spikelets differing from the sessile, glume I of the sessile spikelets deeply channelled. 2. A. pumilus
- CC. Peduncle of racemes not enclosed in spathiform leaf-sheath.
    - Racemes many, fascicled or panicled, glume I of sessile spikelets glabrous and pitted. 3. A. pertusus.
    - Racemes many and whorled in the panicle; glume I of sessile spikelets muricate on the margins. 4. A. squarrosus.

- BB. Racemes of 3 spikelets on the capillary whorled branches of an erect panicle.
    - Pedicels of upper spikelets half as long as the sessile spikelets or longer.
        - Leaves broad [Pg 183]
            - Leaf-sheaths covered densely with bristly hairs. 5. A. asper.
            - Leaf-sheaths covered with soft hairs. 6. A. Wightianus
    - Pedicels of upper spikelets not half as long as the sessile spikelet.
        - Leaves glabrous and narrow 7. A. monticola.
- AA. The lowest one or more sessile spikelets in all racemes, or at least in one or two, differing from all those above.

- o Racemes digitate, rarely solitary, spikelets all alike in form but differing in sex.
    - ▪ Pedicel 1/3 as long as the sessile spikelets; nodes usually glabrous; ligule usually short and membranous. 8. A. caricosus.
    - ▪ Pedicel 1/2 as long as the sessile spikelets; nodes bearded; ligule large and membranous. 9. A. annulatus.
- o Racemes solitary; lower sessile spikelets very unlike the pedicelled or upper spikelets which are cylindric.
    - ▪ Margin of glume 1 of the pedicelled spikelet unequally winged; ligule is a broad truncate membrane. 10. A. contortus.
- o Racemes two, both sessile, or one sessile and the other pedicelled on a peduncle which is more or less sheathed by a proper spathe, divaricate or deflexed.
    - ▪ Leaf base broad and cordate 11. A. Schœnanthus.

N.B.—This genus is now split into several separate genera, each subgenus being raised to the rank of a genus. But in this book the nomenclature adopted in Hooker's Flora of British India is followed.

[Pg 184]

Fig. 148.—Andropogon foveolatus.

[Pg 185]

Andropogon foveolatus, *Del.*

The stems are slender at first, slightly decumbent at the base and then erect, covered at base with silkily villous sheaths, branches freely above before flowering, the lower portion of stems alone being leafy.

The *leaf-sheath* is somewhat scaberulous, partly green and partly purplish, always shorter than the internode. The *ligule* is short, truncate, hyaline and ciliate. *Nodes* are tumid and purplish with a ring of hairs.

The *leaf-blade* is linear, narrow, sometimes even filiform, acuminate slightly cordate at the base, scabrid throughout with a few scattered long bulbous-based hairs near the base to a distance of less than 1/2 inch about it and varies from 2 to 4 inches in length.

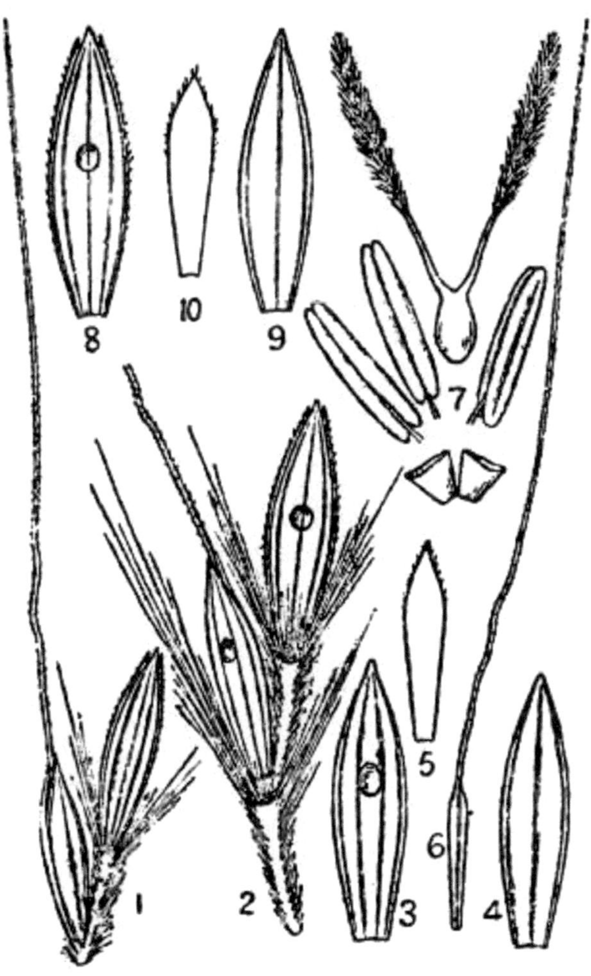

Fig. 149. — Andropogon foveolatus.
1 and 2. Sessile and pedicelled spikelets; 3, 4, 5 and 6. the first, second, third and the fourth glume, respectively, of the sessile spikelet; 7. lodicules, anthers and ovary; 8, 9 and 10. the two glumes and the palea of the pedicelled spikelet.

The *spikes* are solitary, 1 to 1-3/4 inch long exserted far above the small spathiform leaf-sheaths, peduncles are capillary and scaberulous, pedicels and joints are somewhat flattened, and have along both the narrow margins long, white, ascending hairs; callus is short with a ring of short white hairs.

There are two kinds of *spikelets,* sessile and pedicelled, and both are oblong-lanceolate and equal. The *sessile spikelet* consists of *four* [Pg 186] *glumes.* The *first glume* is lanceolate, flat and smooth, keels scabrid with usually a deep dorsal pit, 4-nerved. The *second glume* is lanceolate, acute, as long as the first, 3-nerved. The *third glume* is small, membranous, linear-lanceolate, nerveless. The *fourth glume* is the dilated base of the awn, awn is about 3/4 inch twisted to half its length, scabrid, the lower twisted part dark and the upper pale. There are three *stamens* and two *lodicules. Ovary* has two feathery *stigmas.* The *pedicelled spikelets* have only two glumes and contain three stamens. The *first glume* is oblong-lanceolate, 5-nerved, pitted above the middle, with recurved margins and scabrid keels and nerves. The *second glume* is lanceolate, membranous, hairy at the top, 3-nerved with margins infolded; *palea* is oblanceolate, thinly membranous, nerveless and ciliated at the top; there are three *stamens* and two *lodicules.*

This is a fairly common grass occurring all over the Presidency much liked by cattle and yields plenty of foliage if properly looked after. It grows on all kinds of soils, even laterite.

Distribution. — Throughout India.

[Pg 187]

Fig. 150. — Andropogon pumilus.

[Pg 188]

Andropogon pumilus, *Roxb.*

It is a tufted annual with numerous radiating branches, growing on all directions, bent below and erect above; they vary in length from 6 inches to 18 inches, but sometimes when growing under favourable conditions attain the length of 2-1/2 feet. The stem is slender, green, or pale reddish in the exposed portions and pale in parts covered by sheaths slightly flattened, smooth.

The *leaf-sheaths* are smooth, compressed, distinctly keeled. The *ligule* is a short, truncate, white, glabrous membrane. The *nodes* are glabrous.

The *leaf-blade* is linear, finely acuminate, glabrous, but sometimes somewhat scabrid along the nerves and with scattered long delicate hairs above especially when young, varying in length from 1 to 7 inches and 1/10 to 1/8 inch in breadth.

The *inflorescence* consists of paired spikes with very slender peduncles arising from flattened, glabrous, acuminate spathes, varying in length from 1/2 to 1-1/4 inches. The *spikes* are spreading and one of them always slightly longer than the other, reddish or pale green, 1/2 to 1 inch long; the *rachis* consists of five to eight flat joints broadened at the top and ending in a cup, densely ciliate on both the margins, but hairs on one margin are shorter than those on the other. Each joint bears a sessile and a pedicelled spikelet.

Fig. 151.—Andropogon pumilus.
1. A portion of the spike to show the arrangement of the spikelets; 1.

the first glume of the sessile spikelet; 2. second glume of the sessile spikelet; 3 and 4. third and fourth glumes of the sessile spikelet; 5. anthers, lodicules and the ovary; A, B and C. the three glumes of the pedicelled spikelets.

The *sessile spikelet* is about 3/16 inch with an awn 7/16 inch long. There are four *glumes* in the spikelet. The *first glume* is narrow, linear, membranous, grooved, finely bicuspidate at the apex, with incurved margins and two nerves ending in tubercles below. The *second glume* is a little longer than the first, narrow, lanceolate, boat-shaped, thinly coriaceous with membranous margins, 1-nerved and shortly awned. The *third glume* is about 2/3 of the [Pg 189] second glume in length, and shorter than the first glume, linear-lanceolate, hyaline, nerveless or sometimes very obscurely 2-nerved. The *fourth glume* is narrow linear, hyaline with two very fine lobes at the apex with an awn between, 7/16 inch long. *Palea* is hyaline and very small. *Stamens* are three, *ovary* with two long reddish feathery *stigmas*. *Lodicules* small and cuneate. Grain is long and narrow.

The *pedicelled spikelets* have only three glumes, and are slightly shorter than the sessile ones, pedicel is similar to the joint. The *first glume* is ovate-lanceolate, thinly coriaceous, distinctly many-nerved, acuminate, margins infolded and membranous. The *second glume* is ovate-lanceolate, membranous, glabrous and 3-nerved. The *third glume* is short, oblong-lanceolate, nerveless or faintly 2-nerved. There are three stamens.

This grass is variable in its size. In dry soils such as laterite soils, it is a very small plant not exceeding 9 or 10 inches across its spread. But in good soil and under favourable conditions the plant measures across 5 or 6 feet. Cattle eat the grass before it flowers and do not relish it so much when in flower.

A common grass flourishing all over the Presidency.

Distribution.—Occurs in drier parts throughout India.

[Pg 190]

Fig. 152.—Andropogon pertusus.

[Pg 191]

Andropogon pertusus, *Willd.*

This grass is perennial. Stems are tufted, very slender, widely creeping on all sides, purplish, but the flowering branches are erect or ascending from a geniculate base, leafy at base, the nodes of the creeping branches rooting and bearing tufts of branches which finally become independent plants at each node, the creeping branches vary in length from 1 to 3 feet and the erect ones from 10 to 18 inches or more.

The *leaf-sheaths* are terete or somewhat compressed, glabrous, sometimes ciliated near the node and shorter than the internode. The *ligule* is a truncate membrane, slightly ciliate or not. *Nodes* are bearded.

The *leaf-blades* in the prostrate branches are crowded, short linear-lanceolate, finely acuminate, soft, shortly hairy along the nerves, sparsely ciliate near the rounded base, varying in length from 1 to 2 inches and in breadth 1/8 to 1/4 inch; but on the flowering branches the leaves are longer, sometimes as long as twelve inches with bigger sheaths.

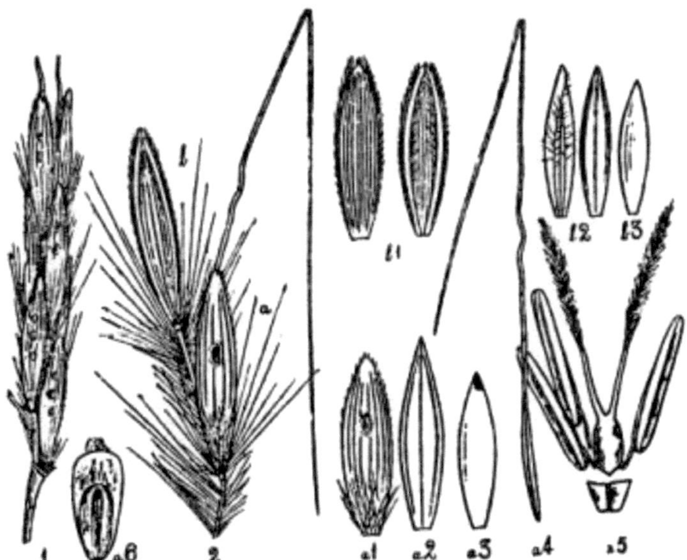

Fig. 153.—Andropogon pertusus.
1. A portion of a spike; 2. a pair of spikelets; a. sessile and b. pedicelled; a-1. first glume; a-2. second glume; a-3. third glume; a-4. fourth glume and awn; a-5. ovary and stamens; a-6. grain; b-1. first

glume of pedicelled spikelet front and back; b-2. second glume front and back; b-3. third glume.

The *inflorescence* consists of three to nine, slender, flexuous, erect, purplish spikes, 1 to 2 inches long, alternately arranged on a thin, long, slender, smooth peduncle of about six inches; *rachis* is slender and the joints and pedicels are densely silky with long hairs.

The *spikelets* are in pairs, one sessile and one-pedicelled, both are equal, purplish or pale. The *sessile spikelet* consists of four glumes [Pg 192] and contains a complete flower and the callus is short and bearded with long hairs. The *first glume* is coriaceous, oblong-lanceolate, acute, truncate or emarginate, slightly hairy, or glabrous with a deep pit above the middle (sometimes with two or three pits also) 7- to 9-nerved with a few long hairs below the middle and with margins infolded and shortly ciliate. The *second glume* is lance-olate-acuminate and finely pointed at the tip and the point project-ing slightly beyond the first glume, 3-nerved or 3- to 5-nerved, membranous, slightly hairy or glabrous, obscurely keeled. The *third glume* is thin, membranous, shorter than the second glume, linear-oblong, subobtuse or acute at the tip and nerveless. The *fourth glume* is the base of the awn and the *awn* is not twisted, bent at about the middle, 1/2 to 2/3 inch long; there is no palea. *Anthers* are three and yellow; *stigmas* purple. The grain is oblong-obovate, slightly trans-parent.

The *pedicelled spikelets* are slightly narrower than the sessile, gen-erally not pitted (though pitted in some plants), and not awned, and each one consists of three glumes only; the pedicel is more than half as long as the sessile spikelets. The *first glume* is slightly hairy, ob-long-lanceolate, acute or obtuse, ciliate at the margins, 7- to 9-, or 13-nerved, generally without pits, but occasionally with one, two or three pits; the keels are ciliolate throughout the length. The *second glume* is membranous, ovate-lanceolate, acute, with incurved mar-gins, 5-nerved. The *third glume* is hyaline, linear-oblong, glabrous and thinly ciliate at the tip or not with or without stamens.

This is an excellent fodder grass and it grows quickly and stands cutting very well. Cattle eat this grass very well.

Distribution. —This grass is found all over India in the plains or lower elevations of hills.

Andropogon squarrosus, *L.f.*

(<u>*Vetiveria*</u> *zizanioides*.)

This is a densely tufted perennial grass with branching root-stocks and spongy aromatic roots.

The stems are leafy, with equitant, hard, leaf-sheaths at the base, smooth and polished, solid, 2 to 3-1/2 feet high.

The *leaf-sheaths* are smooth, coriaceous, glabrous, keeled and compressed. The *ligule* is a very short membrane.

Leaf-blades are narrowly linear, erect, strongly keeled and flat, acuminate, glabrous both above and below, very much narrower than the sheath at the base, 1 to 2 feet by 1/3 to 3/4 inch.

The *panicle* is conical, erect with branches, fascicled, varying in length from 4 to 12 inches. The *spikes* consist of both sessile and pedicelled spikelets, that are either grey, green, or purplish.

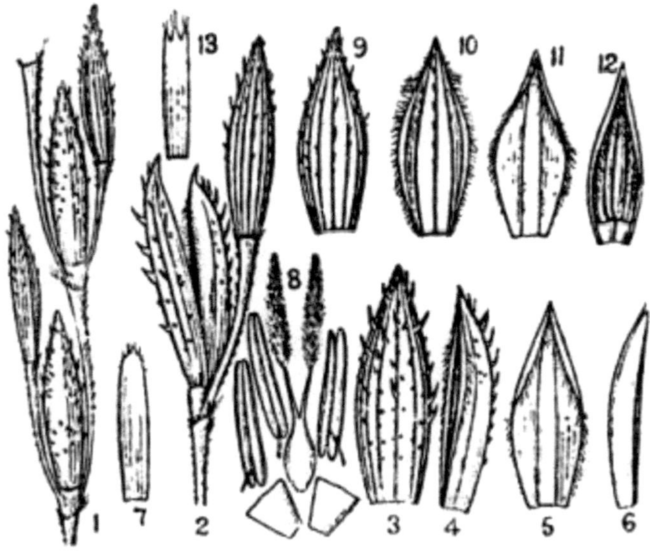

Fig 154.—Andropogon squarrosus. 1. A portion of a branch; 2. a sessile and a pedicelled spikelet; 3, 4, 5 and 6. the first, second, third and the fourth glume, respectively, of the sessile spikelet; 7. palea of the fourth glume; 8. lodicules, stamens and the ovary; 9, 10, 11 and 12. glumes of the pedicelled spikelet; 13. palea of the fourth glume.

The *sessile spikelets* are about 1/6 inch long, lanceolate and with a shortly bearded callus. The *first glume* is ovate-oblong, thickly coriaceous, obscurely 2- to 4-nerved (occasionally 5- to 7-nerved), acute, dorsally flat, with incurved margins and with two rows of tubercle-based minute prickles or mere excrescences at the sides. The *second glume* is as long as the first, oblong, coriaceous, keeled, with hyaline and ciliolate margins, 1-nerved (sometimes 3-nerved, marginal faint), and with minute prickles on the keel. The *third glume* is broadly oblong, hyaline, nerveless or rarely with two obscure veins ciliolate at the margins and acute or acuminate. The *fourth glume* is shorter than the third, linear-oblong, mucronate or very shortly awned at the apex, paleate; *palea* about two-thirds the length of the glume, lanceolate. *Lodicules* are two, quadrate and conspicuous [Pg 194] though small. *Styles* and *stigmas* short. *Stamens* are three with yellow anthers. *Stigmas* are purple.

The *pedicelled spikelets* are similar to the sessile ones, but are slightly smaller and the prickles are less prominent. The *fourth glume* has no mucro or awn and has three stamens.

This grass is fairly abundant in moist situations, in the margins of tanks and in tankbeds in the Coromandel districts, but in other inland districts it is not so common. In some places it seems to be cultivated. This is the *khus-khus* grass.

Distribution.—Throughout the plains and lower hills of India, Burma and Ceylon, also said to occur in Java and Tropical Africa.

[Pg 195]

Andropogon asper, *Heyne.*

(*Chrysopogon asper*, Heyne.)

This is a tufted perennial grass. Stems are stout below with distichous leaves and very slender above, 2 to 3-1/2 feet long.

The *leaf-sheaths* are distichous and towards the base of the stem are 1/2 inch broad, compressed, keeled and with scattered tubercle-based hairs. The *ligule* is a short membrane fringed with close set hairs.

Fig. 155. — Andropogon asper.
Leafy shoot, a bit of the stem with leaf-sheaths and a bit of the leaf.

The *leaf-blades* are broad, distinctly linear, acute or acuminate, coriaceous, glabrous or softly hairy on both the surfaces, with a slender midrib which bears short stiff tubercle-based hairs all along, and margins with similar hairs, but a few leaves towards the [Pg 196] base are longer, and varying in length from 12 to 18 inches and in breadth from 1/2 to 3/4 inch.

The *panicle* is somewhat narrow, 7 to 8 inches long, branches are very slender, whorled, usually with only one spike consisting of a sessile and two pedicelled spikelets.

The *sessile spikelets* are 1/4 inch long, laterally compressed, with a long callus villous all round, and bisexual. The *first glume* is coriaceous, linear-oblong, strongly compressed above and with a few stiff short bristles beneath the tip. The *second glume* is linear, oblong, coriaceous, with an awn as long as itself or shorter, keeled and with short stiff bristles on the keel and on the sides above the middle. The *third glume* is hyaline, narrow, obtuse, shorter than the second, 2-nerved, ciliate. The *fourth glume* is the linear, hyaline, 3-nerved base of the awn; the *awn* is 1-1/2 to 2-1/2 inches and bent at about the middle.

The *pedicelled spikelets* are about 1/3 inch, narrowly lanceolate, male or neuter and with short rusty hairs on both the margins of the pedicel and a semi-circular tip. The *first glume* is thin, 2-toothed or not at the tip, awned, *awn* being as long as itself or longer, 7-nerved, ciliate at the sides from base to tip; the nerves are either equidistant or the lateral nerves nearer the margin. The *second glume* is lanceolate-acuminate, not awned, 3-nerved, margins hyaline, and ciliolate. The *third glume* is hyaline, linear-oblong, 2-nerved, ciliolate. The *fourth glume* is linear or linear-lanceolate, hyaline, nerveless or 1-nerved.

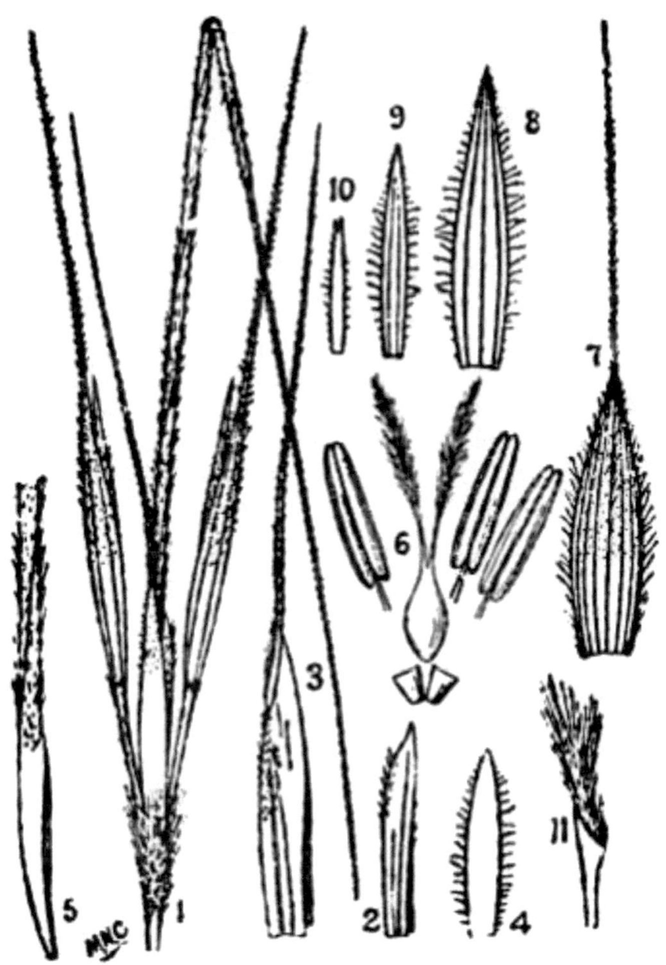

Fig. 156. — Andropogon asper.
1. Spike; 2, 3, 4 and 5. the first, second, third and the fourth glume, respectively, of the sessile spikelet; 6. the ovary, lodicules and stamens; 7, 8, 9 and 10. the first, second, third and the fourth glume, respectively, of the pedicelled spikelet; 11. callus of the spike.

This grass grows abundantly on the sides of the Kambakkam Drug, Chingleput district, and in Penchalkonda, Nellore district, and seems to be an endemic species. It is usually confined to the hill sides and not found in the plains. This grass is very closely allied to *Andropogon Wightianus* and it differs from it only in the general habit of the plant and in having bristles on the leaf-sheaths. On the whole this is a coarser and larger plant than *A. Wightianus*.

Distribution.—Kambakkam Drug in the Chingleput district and Penchalkonda in Nellore district.

[Pg 197]

Andropogon Wightianus, *Steud.*

(*Chrysopogon Wightianus*, Nees.)

This is a perennial. Stems are erect or ascending from a creeping root-stock, varying in height from 2 to 3 feet.

The *leaf-sheath* is flattened, softly hairy or glabrous, often ciliated near the mouth. The *ligule* is a fringe of very short hairs.

The *leaf-blade* is narrowly or rarely broadly linear, obtuse or acute and abruptly mucronate, or narrowly drawn into a point glabrous or pubescent, margins shortly ciliate.

The *panicle* is narrow, 3 to 6 inches long, peduncle smooth below but thinly pubescent above, lower branches long, few in a whorl; rachis is very slender, angular, glabrous or hairy. The *spikes* are solitary and each one consists of one sessile and two pedicelled spikelets. The callus is long and densely bearded with brown hairs.

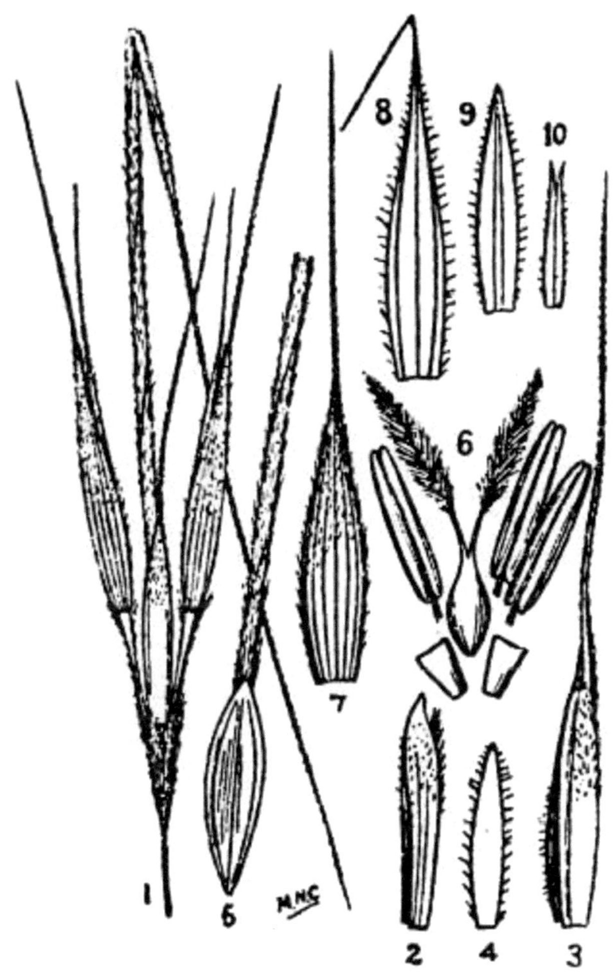

Fig. 157.—Andropogon Wightianus.
1. A spike; 2, 3, 4 and 5. the first, second, third and the fourth glume, respectively, of the sessile spikelet; 6. lodicules, stamens and the ovary; 7, 8, 9 and 10. the first, second, third and the fourth glume, respectively, of the pedicelled spikelet.

Sessile spikelets are bisexual, sub-cylindric about 1/4 inch long. There are four *glumes*. The *first glume* is chartaceous, laterally compressed, obscurely 4-nerved, glabrous below, hispid near the apex, minutely 2-toothed or not at the apex, not awned or rarely with a short awn. The *second glume* is chartaceous, distinctly awned, the *awn* being as long as the glume or longer, hispid above and at the sides also. The *third glume* is hyaline, linear-oblong, 2-nerved ciliate. The *fourth glume* is narrow with hyaline margins, [Pg 198] with an *awn* 2 to 3 inches long; *awn* is hispid below, twisted and geniculate at and less hairy above the middle. Stamens are three. Styles are two and feathery. Lodicules are very small.

Pedicelled spikelets are male or neuter, flattened, hairy, rarely glabrous. The pedicels are half as long or slightly longer than the sessile spikelet, truncate or semi-circular at the top, and with brown villous hairs along the margin. There are four *glumes*. The *first glume* is about 3/8 inch, ciliate, along the inflexed margin, 7-nerved, awned; *awn* equal to or longer than the glume. The *second glume* is as long as the first, shortly awned or acuminate, 3-nerved, ciliate. The *third glume* is hyaline, oblong, 2-nerved, sparsely ciliate. The *fourth glume* is narrow, ciliate, nerveless or rarely 1-nerved, erose or bifid at the top. *Anthers* three or more.

This grass grows on the plains as well as on the hills. It is very closely allied to *Andropogon asper, Heyne,* and it is very difficult to distinguish them. *Andropogon Wightianus* is somewhat smaller compared with *Andropogon asper,* and the tubercle-based bristles on the leaf-sheaths, so characteristic of *A. asper,* is absent.

Distribution.—Madras, Chingleput district, Kodaikānal and the Nilgiris.

[Pg 199]

Andropogon monticola, *Schult.*

(Chrysopogon monticola.)

This is a perennial grass.

The stems are usually slender, densely tufted, erect, simple, or branched, leafy especially at the base, varying in height from 1 to 3 feet.

270

The *leaf-sheaths* are sparsely hairy or glabrous, the lower some-what compressed and the upper terete. The *ligule* is a short, ciliated membrane. The *nodes* are glabrous.

The *leaf-blade* is narrow, linear, acute, rigid, flat, glaucous, smooth or scaberulous, with margins scabrid and ciliated with tubercle-based hairs especially towards the base, and varying in length from 2 to 15 inches.

The *inflorescence* is an open panicle, ovate or oblong, varying in length from 2 to 5 inches; the *rachis* is slender, smooth or scaberulous, the branches are capillary, whorled and spreading, tip oblique, bearded and bearing a single sessile and two pedicellate spikelets.

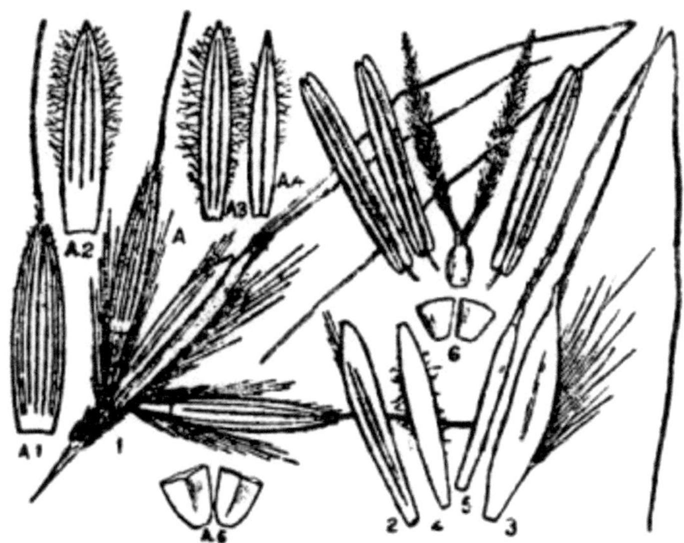

Fig. 158. — Andropogon monticola.
1. Sessile and pedicellate spikelets; 2, 3, 4 and 5. the first, second, third and the fourth glume, respectively, of the sessile spikelet; 6. anthers, ovary and lodicules; A-1, A-2, A-3 and A-4. the glumes of the pedicelled spikelet; A-5. lodicules of the pedicelled spikelet.

The *sessile spikelets* are bisexual, about 1/4 inch or less, with a long callus bearded on one side with long rusty hairs. There are four *glumes* in the spikelet. The *first glume* is chartaceous, linear, complicate, 2-toothed at the tip and with short bristles towards the apex, 4-veined. The *second glume* is chartaceous, ovate-lanceolate, much broader than the first, ciliate with long rufous bristles on the keel, shortly toothed at the apex with an *awn* about 1/3 of an inch and with broadly hyaline margins. The *third glume* is hyaline, narrow-oblong, ciliate and obtuse. The *fourth glume* is narrow, oblong, hyaline with an *awn* nearly an inch long. There are [Pg 200] three *stamens* and two *lodicules*. The *stigmas* are long and feathery.

The *pedicelled spikelets* are as long as the sessile and the pedicels are flattened and with long rufous hairs on both the margins. There are four *glumes*. The *first glume* is lanceolate, acute and awned between two teeth, 7-nerved and scaberulous. The *second glume* is lanceolate, acuminate, with thinly ciliate hyaline margins, 3-nerved. The *third glume* is shorter than the second, narrow, hyaline, ciliate at the margins, 2-nerved. The *fourth glume* also is small, hyaline, ciliate, and 1-nerved. There are three *stamens* and two *lodicules*.

This grass is found growing all over the Presidency on the plains and even on low hills. It grows into a tall plant in rich soils and remains stunted in poor, dry and rocky soils. Cattle eat this grass.

Distribution. — Throughout India and Ceylon and in Africa.

[Pg 201]

Andropogon caricosus, *L.*

This is a perennial grass more or less tufted in habit and closely allied to *Andropogon annulatus*, Forsk.

Stems are erect or decumbent below or ascending from a creeping base, rooting at the nodes, smooth, glabrous and much branched, varying in height, from 1 to 2 feet; branches are short, slender and sometimes even capillary, with *nodes* bearded or not in branches ending in solitary spikes, and completely glabrous when they end in binate spikes.

The leaf-sheaths are glabrous, rather compressed, striate, shorter than the internodes. *Ligule* is membranous, short, very finely ciliolate or not.

The leaf-blade is linear, finely acuminate, sparsely hairy, sometimes with tubercle-based hairs, becoming glabrous when old with scaberulous margins 2 to 8 inches by 1/10 to 1/6 inch, base rounded mostly with a few long hairs.

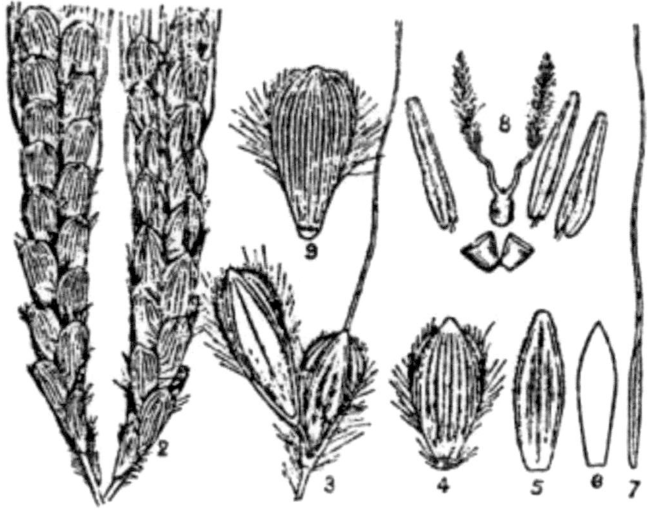

Fig. 159. — Andropogon caricosus.
1 and 2. Front and back view of a bit of spike; 3. a sessile and a pedicelled spikelet; 4, 5 and 6. the first, second and the third glume, respectively, of the sessile spikelet; 7. awn representing the fourth glume; 8. stamens, lodicules and the ovary; 9. the first glume of the pedicelled spikelet.

The spikes are either binate or solitary varying in length from 1 to 2 inches, joints and pedicels about 1/3 as long as the sessile spikelets, slightly angular or flat, ciliate along one side with white hairs; peduncle is slender, pale or purple, pubescent or glabrous just below the spike.

The spikelets are about 1/8 inch, imbricate, a sessile and a stalked one from the top of each joint, greenish or purple. The *sessile spikelet* contains a bisexual flower and consists of four glumes. The callus is short, and shortly hairy below. The *first glume* is somewhat chartaceous, obovate-oblong, obtuse or truncate, 7- to 11-nerved, margin slightly folded, keel shortly rigidly ciliate towards the apex, and thinly ciliate below, dorsal surfaces sparsely hairy [Pg 202] below the middle. The *second glume* is chartaceous, ovate-lanceolate, acute, equal to or slightly longer than the first glume but narrower, 3-nerved, margin infolded, thinly shortly ciliate, dorsally glabrous, shining. The *third glume* is hyaline, ovate-oblong, acute, nerveless, margins sparsely ciliate or not. The *fourth glume* is the base of the awn, 3/4 to 1 inch, scaberulous. *Stamens* are three with yellow or purple tinged anthers, *ovary* oblong with two feathery *stigmas. Lodicules* are two, cuneate.

The *pedicelled spikelets* are either male or neuter and consist of four *glumes*. The *first glume* is chartaceous, obovate-oblong, obtuse, many-nerved (thirteen or more), thinly ciliate with long hairs and with a few rigid short hairs towards the apex; margins are slightly infolded, dorsally sparsely hairy without. The *second glume* is membranous, ovate-lanceolate, acute, 3-nerved (occasionally 4-nerved), margins are thinly ciliate and infolded. The *third glume* is hyaline, nerveless and ciliate. The *fourth glume* is hyaline, nerveless, linear and oblong, glabrous, small, the apex is narrowed and deeply bifid. There are three *stamens* and two *lodicules*.

This is a common grass flourishing on the bunds of paddy fields and in sheltered places where there is sufficient moisture in the soil. But this is less common than *A. annulatus*, Forsk. In black cotton soil at Bantanahal in Bellary district it grows to a height of 4 or 5 feet.

Distribution. — Plains and low hills throughout India and Ceylon.

[Pg 203]

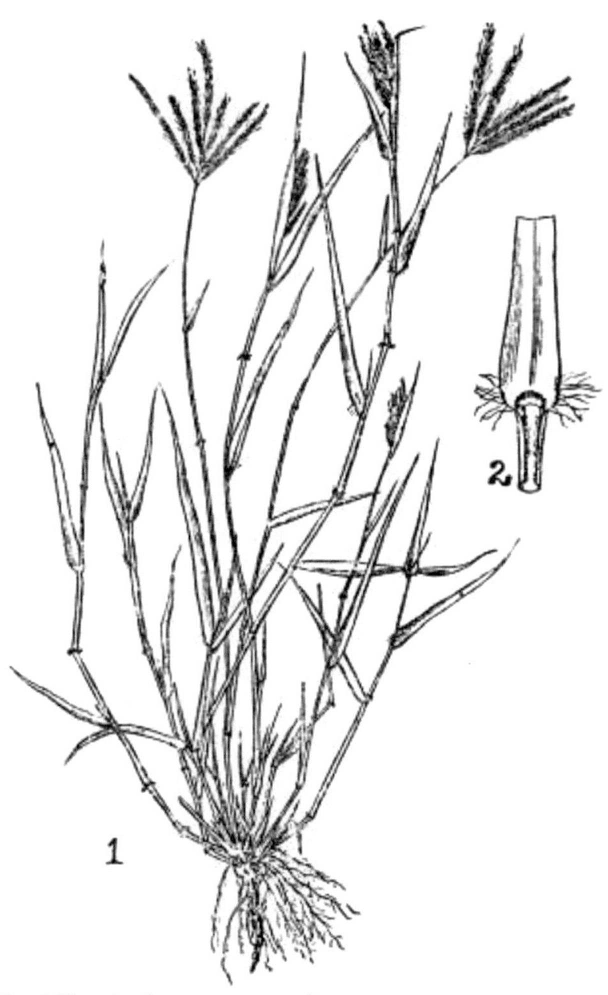

Fig. 160. — Andropogon annulatus.
1. Full plant; 2. base of the leaf and ligule.

[Pg 204]

Andropogon annulatus, *Forsk.*

This is a densely tufted perennial grass.

The main stem is underground, rhizomiferous, and covered with scale leaves; branches are many arising in tufts, leafy, procumbent at base and afterwards geniculately ascending and ending in inflorescence, occasionally rooting at the nodes and varying in length from 2 to 3 feet. The internodes vary from 1-1/2 to 4 inches, pale or purplish, slightly flattened, smooth and glabrous.

The *leaf-sheath* is terete, glabrous, shining, green or purplish, closed, with margins where separate ciliated and profusely so at the tip especially the outer or both. The *ligule* is membranous truncate, glabrous, about 1/16 inch in height. *Nodes* are purple and softly villous.

Fig. 161.—Andropogon annulatus.
1. Front and back views of a portion of the spike; 2. a sessile and a pedicelled spikelet; 3, 4, 5 and 6. the first, second, third and the fourth glume, respectively, of the sessile spikelet; 7. the ovary, stamens and lodicules; 8, 9 and 10. the glumes of the pedicelled spikelet.

The *leaf-blade* is linear-lanceolate, acuminate, scabrid, sparsely hairy, becoming glabrous except at the base and with tubercle-based hairs on the upper surface.

The spikes vary in number from two to nine, erect or slightly spreading, subdigitately fascicled, pale when young and pinkish or brown when old, varying in length from 1 to 2-1/2 inches. The stalk of the whole inflorescence is long, slender, smooth and glabrous. The *peduncle* of the spikes is from 1/8 to 1/6 of an inch long, thin, slender, glabrous with swollen bases and with a ring of hairs at the node. *Joints* of the *rachis* and the *pedicels* are slightly flattened, ciliated along the narrow edges; the *pedicels* of the stalked spikelets are half as long as the sessile spikelets. The spikelets are one sessile and one pedicelled and imbricating on the rachis.

[Pg 205]

The *sessile spikelet* is as long as the stalked or a little less, with a thick callus, shortly bearded at the base or sometimes glabrous and consists of four *glumes*. The *first glume* is elliptic-oblong or oblong, obtuse or truncate, irregularly 2- or 3-toothed, 5- to 9-nerved, sparsely villous with long hairs and margins slightly infolded. The *second glume* is smaller than the first glume, acute, membranous, 3-nerved and keeled, the margins are ciliate and infolded. The *third glume* is hyaline, linear, acute, or obtuse, nerveless sparsely hairy at the tip, very much shorter than the second glume. The *fourth glume* is an *awn* with a linear hyaline base, erect, about an inch long. *Stamens* are three, *ovary* is oblong with two feathery, dark purple *stigmas*. *Lodicules* are two, cuneate.

The *pedicelled spikelets* are male and consist of only three glumes. The *first glume* is elliptic, oblong, irregularly obtuse, about 11-nerved, margins slightly infolded with long pilose hairs throughout, more along the margin. The *second glume* is a little smaller, 3-nerved, sparsely hairy only along the marginal nerves, folded inwards, and slightly keeled. The *third glume* is shorter than the second, hyaline, nerveless, narrow-lanceolate, acute; *stamens* are three, with green anthers, purple-dotted. *Lodicules* are two, broad and cuneate.

This grass is found flourishing all over India and grows in cultivated fields and gardens and likes sheltered places. This yields a considerable amount of fodder and stands cutting well.

Fig. 162.—Andropogon contortus.
[Pg 207]

Andropogon contortus, *L.*

(Heteropogon contortus, Beauv.)

This is a tufted perennial.

The stems are erect or slightly decumbent below, slender, rather compressed towards the base, leafy at the base, simple or branched, densely tufted and varying in length from 1 to 3 or 4 feet.

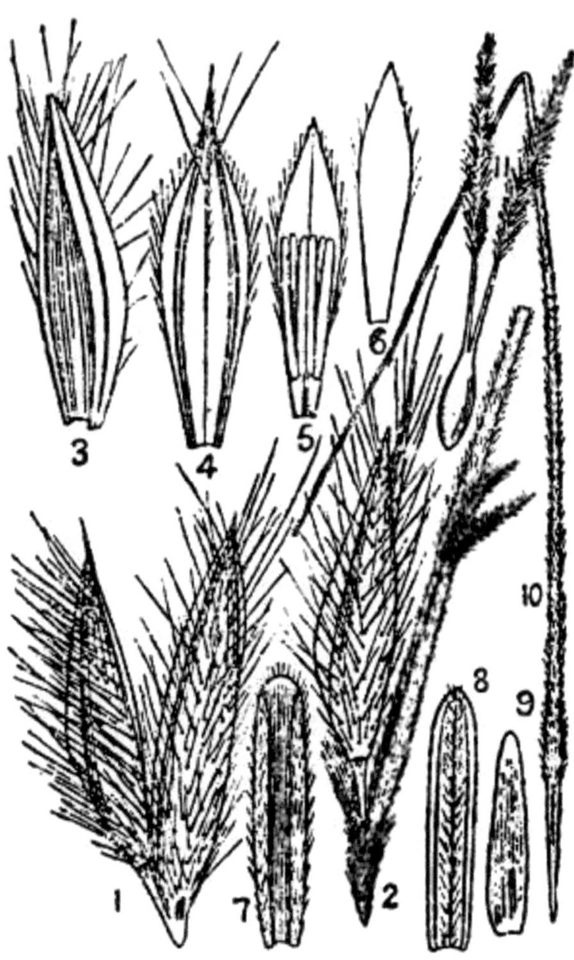

Fig. 163.—Andropogon contortus.
1. Lower pair of sessile and pedicelled spikelets; 2. upper pair of sessile and pedicelled spikelets; 3, 4, 5 and 6. the first, second, third and the fourth glume, respectively, of pedicelled spikelets; 7, 8, 9 and 10. the first, second, third and the fourth glume, respectively, of the sessile spikelet; 11. ovary.

The *leaf-sheath* is smooth or sparsely hairy, compressed and shortly auricled or not at the mouth. The *ligule* is short, truncate and ciliolate.

The *leaf-blades* are linear, acute or abruptly acuminate, flat, rigid, sparingly ciliate above, with tubercle-based hairs towards the base, scaberulous throughout, and 2 to 12 inches long or more, 1/10 to 1/5 inch broad.

The *inflorescence* consists of a solitary spike with closely imbricating spikelets.

The *spikelets* are all on one side, and the lower two to six pairs of pedicelled and sessile spikelets are all males. The *sessile spikelets* are all female and awned, except the few lower which are male [Pg 208] and awnless, 1/4 inch long. The *callus* is long, acute, bearded with reddish-brown hairs. There are four *glumes* in the spikelet. The *first glume* is narrow, linear-oblong, truncate or rounded, somewhat brown, many-nerved, hispid, with incurved margins and membranous tip. The *second glume* is linear, obtuse, coriaceous, dark-brown, hispidulous, 3-nerved with incurved margins. The *third glume* is oblong, hyaline, thin, nerveless, short and truncate. The *fourth glume* is reduced to an awn, 3 inches or more in length. The *ovary* is linear with two long *stigmus*.

The *pedicelled spikelets* are somewhat longer than the sessile 1/3 to 1/2 inch, with very short pedicels. The *first glume* is lanceolate, obliquely twisted, hispid at the back with long bulbous-based hairs, margins more or less unequally winged. The *second glume* is oblong lanceolate, acuminate, 5-nerved, thinly ciliate with hyaline margins. The *third glume* is oblong, hyaline, 1-nerved and ciliate. The *fourth glume* is obovate-oblong or oblong, hyaline, ciliate, nerveless. There are three *stamens*.

This grass though coarse forms very good hay if cut before it flowers. The only objection against this grass is the presence of the troublesome awns which get twisted together like the strands of a rope. This is the *spear grass* of the Anglo-Indians. It grows all over the Presidency and is a troublesome weed when in flower.

Distribution. — All over the Presidency and India. Common in all tropical countries.

Andropogon Schoenanthus, *L. Var. cæsius.*

(Cymbopogon cæsius, Stapf.)

This is a perennial grass with stout or slender, erect stems rising from a woody base, leafy upward, simple or branched.

The *leaf-sheath* is smooth and glabrous. The *ligule* is an oblong-ovate membrane. *Nodes* are glabrous.

The *leaf-blade* is long, narrow or broad, narrowly linear-lanceolate, finely acuminate, glaucous especially beneath, thinly coriaceous, glabrous on both the surfaces, base rounded or cordate and amplexicaul, 6 to 10 inches by 1/6 to 1/3 inch.

The *panicle* is elongate, leafy, narrow, dense or interrupted, compound or decompound, 1 to 2 feet long; bracts are lanceolate, spathiform, finely acuminate, glabrous, varying in length from 1 to 1-1/2 inches, and with hyaline margins; the proper bracts are as long as the spikes or longer.

Fig. 164.—Andropogon Schoenanthus.
1. A sessile and two pedicelled spikelets; 2, 3, 4 and 5. the first, second, third and fourth glume of the sessile spikelet, respectively; 6. ovary; 7, 8 and 9. the glumes of the pedicelled spikelets in order.

The *spikes* are unequal, 1/2 to 2/3 inch long, one 3- to 4-jointed and the other 4- to 6-jointed; the joints and pedicels are narrowly clavate, half as long as the sessile spikelets, tips dilated and toothed, margins villously ciliate, with long hairs.

The *spikelets* are binate, one sessile and the other pedicelled.

The *sessile spikelets* in the upper part of the spike are bisexual, lanceolate, 1/6 inch long and those in the lower part of the spike are shorter, obtuse, male. The callus is short and bearded. There are four *glumes*. The *first glume* is ovate or obovate-oblong, dorsally flat or nearly so, with a deep narrow-longitudinal median furrow usually below the middle and answering to a ridge on the ventral face, obtuse or 2-toothed at the apex, margined above the middle, with a hyaline, narrow, finely denticulate wing, 2-nerved or nerveless. The *second glume* is lanceolate, cymbiform, acute or acuminate, 3-nerved, margins hyaline, ciliate, as long as the first [Pg 210] chartaceous and the keel with a serrulate wing above the middle. The *third glume* is linear oblong, hyaline, obtuse, ciliate, nerveless. The *fourth glume* is the narrowly winged 2-lobed base of the awn, lobes are lanceolate erect and *palea* of the fourth glume is minute. *Lodicules* are cuneate. *Stamens* are three.

The *pedicelled spikelets* are oblong-lanceolate, acute or obtuse, glabrous and male. There are three *glumes*. The *first glume* is glabrous or rarely puberulous, margins incurved, obtuse, 9- to 11-nerved. The *second glume* is ovate, acute, 3-nerved. The *third glume* is oblong or linear-oblong, hyaline, apex rounded, ciliate and faintly 2-nerved.

This grass grows all over the Presidency in open dry situations and is very widely distributed.

Distribution.—Throughout India—westward to tropical Africa.

[Pg 211]

29. Anthistiria, *L. f.*

(Themeda, Forsk.)

These are tall grasses, annual or perennial. Leaves are usually long and narrow. The inflorescence consists of racemes or panicles of fascicled spikes in the axils of spathiform bracts. The spikelets vary in number from six to eleven in a cluster, the four lowest being male or neuter, and forming an involucre with whorled or superposed pairs round either 1-sessile bisexual spikelet with two pedicelled spikelets or two superposed bisexual, the lower with one pedicelled, the upper with two.

The involucral spikelets are male or neuter, the largest, and consist of three glumes. The first glume is oblong, lanceolate, dorsally flattened, many-nerved, margins narrowly incurved and keels narrowly winged. The second glume is membranous, lanceolate, acute, 3-nerved, with ciliate margins. The third glume is hyaline, smaller than the second, 1-nerved or this glume may be absent, stamens have large anthers. The pedicelled spikelets are similar to the involucral in every respect but smaller, male or neuter, but the first glume is not winged on the keels. The bisexual (or female) spikelets are smaller than the involucrant spikelets, linear-oblong, subterete, obtuse with a rigidly bearded callus. There are four glumes in the spikelet. The first glume is terete, or dorsally compressed or channelled, coriaceous and at length hardened, margins incurved, dark brown to almost black when old. The second glume is as long as the first, linear, dorsally chartaceous, with broadly incurved membranous margins, 3-nerved. The third glume is very small, hyaline, 1-nerved, epaleate. The fourth glume is the flattened base of the awn, epaleate. The lodicules are two, cuneate. Anthers are rather small. Styles are laterally or terminally exserted. Grain is narrow, obovoid, biconvex, with two grooves on the anterior side and with a long embryo.

[Pg 212]

Anthistiria tremula, *Nees* .

This is an annual or perennial. Stems are stout or slender, erect or ascending from a creeping root-stock, simple or branched, 1 to 4 feet.

284

The *leaf-sheath* is smooth, compressed. The *ligule* is a narrow membrane.

The *leaf-blade* is linear-lanceolate, rigid, erect, acuminate with a setaceous tip, nearly smooth, varying in length from 6 to 20 inches and in breadth from 1/6 to 2/3 inch.

The *inflorescence* is an elongate panicle, 1 to 2 feet long, consisting of rather distant fascicles of spikes and bracts on capillary, flexuous peduncles; the spikes are sub-flabelliform or sub-globose, 1/2 to 1-1/2 inches broad, sometimes reduced to a few spikelets and bracts, the outer bracts are longer than the fascicles, 1 to 1-1/2 inches long, glabrous or hairy with ordinary or tubercle-based hairs; proper bracts are lanceolate, acute, compressed, glabrous or hairy with membranous margins.

Fig. 165. — Anthistiria tremula.
1. Fascicles of three spikes with the outer bracts and proper bracts;
2. a spike without its proper bract; 3. the pedicelled and the bisexual spikelets without the involucral spikelets; 4, the first glume of the involucral spikelet with one wing only; 4a. the first glume of the involucral spikelet with wings to both the keels; 5 and 6. the second and the third glume of the involucral spikelet; 7, 8 and 9. the glumes of the bisexual spikelet; 10, 11, 12 and 13. glumes of the bisexual spikelet; 14. ovary.

The *involucral spikelets* are the longest, in contiguous superposed pairs, about 1/2 inch long, and the rachis of the spike is produced beyond these spikelets. There are three *glumes.* The *first glume* is linear-lanceolate, acute, covered with long, often tubercle-based hairs, many-nerved, margins narrowly incurved, and with narrow wings, on both the keels in one of each of the pairs of spikelets and on one keel only in the other of each of these pairs. The *second glume* is oblong-lanceolate, acute, margins thin and membranous, inflexed, ciliate above the middle, 3-nerved. The *third glume* is as long as the second, hyaline, very narrowly linear, 1-nerved. *Stamens* are three and the *lodicules* are cuneate.

The *pedicelled spikelets* are usually smaller than the involucral spikelets and similar to them. The *first glume* is winged on one side in the lowest spikelet and without wings in the others.

[Pg 213]

The *bisexual or (female) spikelets* are linear-oblong, obtuse, and the callus with reddish hairs. The *first glume* is scabrid, deeply channelled at the back, nerveless, narrowly truncate at the tip, and hispid near the apex. The *second glume* is as long as the first, linear, hyaline, 3-nerved, chartaceous at the back with the sides membranous and incurved. The *third glume* is small, hyaline, 1-nerved and epaleate. The *fourth glume* is the narrowed base of the awn which is 1/2 inch long.

This grass is very common in marshes and in wet low-lying places on the hills and occurs also in the plains in Malabar and South Kanara.

Distribution.—The Deccan Peninsula, from the Konkan and Central Provinces southward, and Ceylon.

[Pg 214]

30. Iseilema, *Hack.*

These grasses are either annual or perennial, with slender freely branching stems. The inflorescence is a panicle consisting of groups of dissimilar spikelets with compressed, boat-shaped spathes on peduncles. Spikelets are of two kinds, sessile and pedicelled. Each peduncle bears 4-pedicelled male or neuter spikelets in a regular whorl forming an involucel around 1 or 2 sessile bisexual spikelets and 2- or 3-pedicelled male spikelets. Involucral spikelets have 3 or 2 glumes, the first two glumes are somewhat similar, the first 3- to 5-nerved and the second 3-nerved, the third glume is one nerved and hyaline. Lodicules are cuneate and retuse. Anthers yellow dotted or tinged violet. Pedicelled spikelets inside the involucral similar to those of the involucral. Sessile spikelets are bisexual or sometimes female, 4-glumed and awned.

KEY TO THE SPECIES.

- Panicle slender, lax; involucral spikelets 1/6 inch; pedicel slender, terete 1. I. laxum.
- Panicle crowded, leafy; involucral spikelets 1/6 inch or more, very strongly nerved; pedicel harder, firmer and flattened 2. I. anthephoroides.

[Pg 215]

Fig. 166. — Iseilema laxum.

[Pg 216]

Iseilema laxum, *Hack.*

It is a tufted perennial grass with a stout, short, creeping root-stock. Stems are slender, branched, ascending, 6 to 24 inches long.

288

The *leaf-sheaths* are somewhat loose, glabrous. The *ligule* is a short-ly ciliate membrane.

The *leaf-blade* is linear, obtuse, glabrous and ciliate near the base, 2 to 6 inches long. The leaf-blades in the upper portions of the branches are smaller.

Fig. 167.—Iseilema laxum.
1. A cluster of spikelets with spathes. 2. a cluster consisting of the involucral spikelets and three inner spikelets; 3. the inner spikelets consisting of one sessile female or bisexual and 2-pedicelled male spikelets; 4, 5 and 6. the first, second and the third glume, respectively, of the involucral spikelet.

The *inflorescence* is a narrow long panicle bearing clusters of spikelets with spathes on slender peduncles, the outer spathes are narrow-lanceolate, glabrous or with a few hairs near the margin, 1/4 to 1 inch long; inner spathes are lanceolate, smaller with membranous margins. Each cluster consists of an involucel of 4 pedicelled spikelets forming a true whorl around 2 pedicelled and 1 sessile spikelets or 3 pedicelled and 2 sessile spikelets. The involucral spikelets are male, oblong-lanceolate, acute, with short flattened pedicels, bearded at the base, and have three glumes. The *first glume* is oblong-lanceolate, acute, 5- to 7-nerved and ciliate. The *second*

289

glume is oblong-lanceolate, acuminate, equal or slightly shorter than the first, glabrous, 3-nerved and with infolded margins. The *third glume* is hyaline, linear, short, irregularly toothed at the apex. The inner pedicelled spikelets are similar to the involucral spikelets, but the third glume is very narrow, linear. The sessile spikelets are female, rarely bisexual, narrowly lanceolate, 1/5 inch long, glabrous and have four glumes. The *first glume* is lanceolate, chartaceous, truncate or 2-fid at the apex, [Pg 217] faintly 5-nerved, with a few long hairs or glabrous, and with margins scaberulous towards the tip to about one-third the length of the glume. The *second glume* is lanceolate, acuminate, glabrous, sub-chartaceous, 3-nerved. The *third glume* is hyaline, nerveless, apex irregularly cut, short; sometimes this glume is wanting. The *fourth glume* is a very slender awn of about 1/2 inch.

Fig. 168.—Iseilema laxum.
1. Inner spikelets consisting of 2-pedicelled male and two female or bisexual spikelets; 2, 3, 4 and 5. the first, second, third and the fourth glume, respectively, of the sessile spikelets; 6. ovary; 7, 8 and 9. the first, second and the third glume, respectively, of the inner pedicelled spikelet.

This is a widely spread common grass growing in somewhat moist situations. This is the well-known Chengali gaddi of the Telugu districts.

Distribution.—All over Madras and Bombay presidencies.

[Pg 218]

Fig. 169.—Iseilema anthephoroides.

[Pg 219]

Iseilema anthephoroides, *Hack.*

This is a perennial grass closely resembling *Iseilema laxum* in its habit, but shorter, stouter and branching more freely. The leaf is similar to that of *I. laxum* in all its parts.

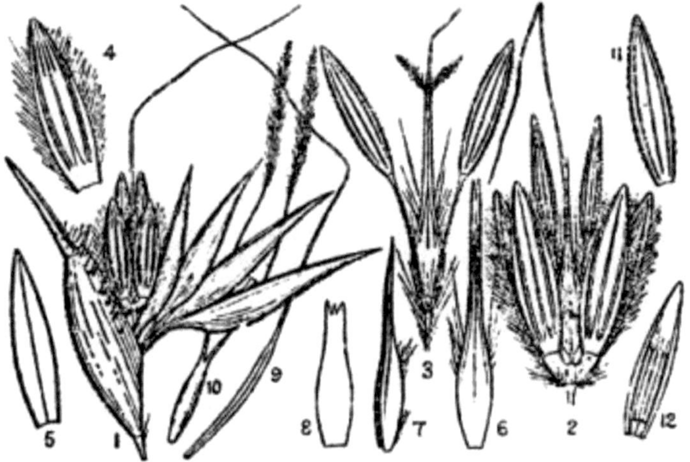

Fig. 170.—Iseilema anthephoroides.
1. A cluster of spikelets with spathes; 2. the involucral and the inner spikelets; 3. the inner spikelets; 4 and 5. the glumes of the involucral spikelets; 6, 7, 8 and 9. the four glumes, respectively, of the sessile spikelet; 10. ovary; 11 and 12. glumes of the inner pedicelled spikelets.

The *pedicelled spikelets* of the involucel have firmer harder, shorter and broader pedicels, thickly bearded and consist of two glumes only. The *first glume* is very strongly 5-nerved, coriaceous, oblong-lanceolate; with scaberulous infolded margins, with long cilia. The *second glume* is lanceolate, thin, 3-nerved, glabrous. The inner *pedicelled spikelets* are similar to the pedicelled spikelets of the involucel. The *sessile spikelet* has four glumes. The *first glume* is elliptic-lanceolate, apex drawn into a long narrow strip ending in two teeth or truncate, sparsely ciliate at the margins about the middle, faintly 3-nerved. The *second glume* is shorter than the first, lanceolate, drawn out into an acuminate point at the apex, hairy at the back.

The *third glume* is hyaline, short, oblong, apex broad and irregularly toothed, nerveless. The *fourth glume* is an awn.

This is very common in the Deccan districts and grows on all kinds of soils. This is a good fodder grass.

Distribution. — Very common in the Ceded districts and Nellore.

[Pg 220]

CHAPTER IX.
Series II — Poaceæ.
TRIBES V AND VI — AGROSTIDEÆ AND CHLORIDEÆ.

The tribe **Agrostideæ** is a very small one. It is represented in South India only by a few genera. The spikelets are usually 1-flowered and the rachilla is jointed at the base just above the empty glumes and it is not produced beyond the flowering glume. There are only three glumes in the spikelet.

- Sub. Tribe 1. **Stipeæ.** — The spikelets are narrow and long, panicles and the flowering glumes are rigid or hard, and awned.
    - The third glume is narrow, long, awn 3-fid. 31. A-ristida.
- Sub. Tribe 2. **Euagrosteæ.** — The spikelets are very small, in open or contracted panicles.
    - The third glume is thin and membranous, awnless. 32. Sporobolus.

Chlorideæ is also a small tribe with about ten genera, most of them being very common in Southern India. The spikelets are uni-laterally biseriate on the rachis which is not jointed at the base. There are one or more flowers in the spikelet, all or only the lowest being bisexual. The rachilla is jointed just above the empty glumes and it is produced or not beyond the flowering glumes. The inflo-rescence consists of spikes, or spiciform racemes, solitary or digi-tate, and in some it is paniculate.

- Rachilla produced beyond the flowering glume.
    - Spikes usually solitary.
        - Spikelets 1- to 2-flowered, pedicelled and in deciduous clusters, awned. 33. Gracilea.
        - Spikelets 1- to 2-flowered, not clustered awned. 34. Enteropogon.

- o   Spikes or spiciform racemes digitate or whorled.
    - ▪   Spikelets 1-flowered and with three glumes, awnless. 35. Cynodon.
- o   Rachilla not produced beyond the flowering glumes.
    - ▪   Spikelets 2- or more-flowered, glumes five or more, awned, upper flowers imperfect. 36. Chloris.
    - ▪   Spikelets 3- to 6-flowered, densely crowded, awnless. 37. Eleusine.
    - ▪   Spikes or spiciform spikes racemed, spikelets 2- to 3-flowered, 4- to 5-glumed, awned. 38. Dinebra.
    - ▪   Spikes panicled, filiform, spikelets very minute one-or more-flowered, glumes awnless. 39. Leptochloa.

[Pg 221]

31. Aristida, *L.*

These are tufted, annual or perennial grasses. Spikelets are panicled, 1-flowered, laterally compressed, with the rachilla jointed above the empty glumes, 3-glumed. The first and the second glumes are narrow, keeled, 1-nerved, awned or not and persistent. The third glume is very narrow, cylindric, coriaceous, convolute, acuminate, 3-nerved, tip produced into a long 3-partite, naked or hairy awn twisted below the branches, with a minute palea which is convolute round the ovary. Lodicules are two, linear or oblong-linear and hyaline. Stamens are three. Styles are distinct. Grain is long, narrow and cylindrical.

KEY TO THE SPECIES.

- •   Awn tripartite from the base and not articulate with the top of the glume, persistent and glabrous.
    - o   Annual.
        - ▪   Glumes I and II not awned.

- - Awn without any column and branched from the base. 1. A. Adscenscionis.

  - - Awn with a short column and with shorter branches. 4. A. mutabilis.
  - o Perennial.
    - - Panicle cylindric, glumes I and II awned; callus with white silky hairs. 2. A. setacea.
    - - Panicle effuse, glumes I and II awned or not; callus naked. 3. A. Hystrix.
- - Awn with a long column, tripartite at the top.
  - o Annual; panicle lax, narrow; glumes I and II awned. 5. A. funiculata.

[Pg 222]

Fig. 171.—Aristida Adscenscionis.

[Pg 223]

Aristida Adscenscionis, *L.*

This grass is usually an annual becoming a perennial under favourable conditions. Stems are slender, sometimes even filiform,

erect, or ascending, simple or branched, varying in length from 9 inches to 3 feet.

The *leaf-sheath* is glabrous, thinly striate. The *ligule* is a row of fine short hairs. *Nodes* are glabrous.

The *leaf-blade* is narrow, linear, tapering to a fine point, convolute in bud, scabrid above and smooth below, with a minutely serrate, very narrow, hyaline margin, 1 to 10 inches long and 1/12 inch broad.

The *inflorescence* is a lax, narrow, subsecund panicle, varying in length from 3 to 12 inches, and with a slender glabrous peduncle; the main rachis is filiform and glabrous; branches are either solitary or binate, unequal; branched either from the middle or the base; *pedicels* are short and capillary.

Fig. 172. — Aristida Adscenscionis.
1. A spikelet; 2. first and second glumes; 3. palea; 4. lodicules, stamens and ovary; 5. third glume with awns; 6. grain.

The *spikelets* are narrow, erect, green, occasionally also purplish, 1/4 to 1/3 inch long exclusive of the awn. There are three *glumes*. The *first glume* is linear-lanceolate, acute, membranous, 1-nerved with a scaberulous keel, 1/16 to 3/16 inch long. The *second glume* is longer than the first, linear-lanceolate, acute, occasionally 2-toothed and apiculate, 1-veined about 1/4 inch long and with a smooth keel. The *third glume* is as long as the second or slightly longer, laterally compressed, 3-nerved, smooth but scaberulous along the keel, awned; there are three scabrid *awns*, varying in length from 1/2 to 3/4 inch, continuous with the glume without a column, not jointed, and the middle awn is longer than the lateral ones; the callus is long, pointed and villous. There is a minute *palea. Lodicules* are two, similar to the palea in size, linear oblong. *Anthers* are yellow dotted with purple. The *ovary* is oblong linear with two white feathery *stigmas.*

Grain is long and linear.

[Pg 224]

This when young is eaten by cattle, but they do not like it when in flower.

Distribution.—Occurs all over the Presidency in the plains and the low hills.

[Pg 225]

Aristida setacea, *Retz.*

This is a tall coarse perennial grass with hard, smooth and polished, stout, erect simple or branched stems, 3 to 4 feet. Roots are stout and wiry.

The *leaf-sheath* is glabrous, cylindrical. The *ligule* is a row of short hairs. The *nodes* are glabrous.

The *leaf-blade* is linear, coriaceous, convolute, glabrous, strongly nerved, 6 to 12 inches long.

The *inflorescence* is a contracted *panicle* varying from 6 to 18 inches with short, erect or subsecund branches.

Fig. 173.—Aristida setacea.
1. The spikelet; 2 and 3. the first and the second glume; 4. the lower portion of the third glume, anther, ovary and the lodicules; 5. palea of the third glume.

The *spikelets* vary from 1/2 to 2/3 inch excluding the awn. There are three *glumes*. The *first glume* is about 3/8 inch long, lanceolate-linear, narrowed into a short awn. The *second glume* is longer than the first, 1-nerved and minutely 2-toothed or notched at the base of the awn. The *third glume* is 5/8 inch long, 3-nerved, nearly smooth. The callus of the third glume is long, densely silkily hairy with three awns not jointed at the base with the glume; *awns* about 1 inch or more. *Lodicules* are ovate-lanceolate, fairly large. Grain is narrow, cylindrical.

This grass grows in open dry situations in many parts of the Presidency.

Distribution. — All over India.

[Pg 226]

Aristida Hystrix, *Linn. f.*

This is a diffuse perennial grass with a creeping root-stock, with fairly stout sometimes proliferous freely branching stems; branches are stiff, erect, inclined or prostrate, varying in length from 6 inches to 2 feet.

The *leaf-sheath* is glabrous and cylindric. The *ligule* is a ridge of close-set hairs. *Nodes* are glabrous.

The *leaf-blades* are quite flat, narrowly lanceolate-linear very finely acuminate, glabrous on both the surfaces but with tufts of hairs on both sides at the base where the blade joins the sheath, prominently nerved; margin is even and without any hyaline border, the blade varies in length from 2 to 9 inches.

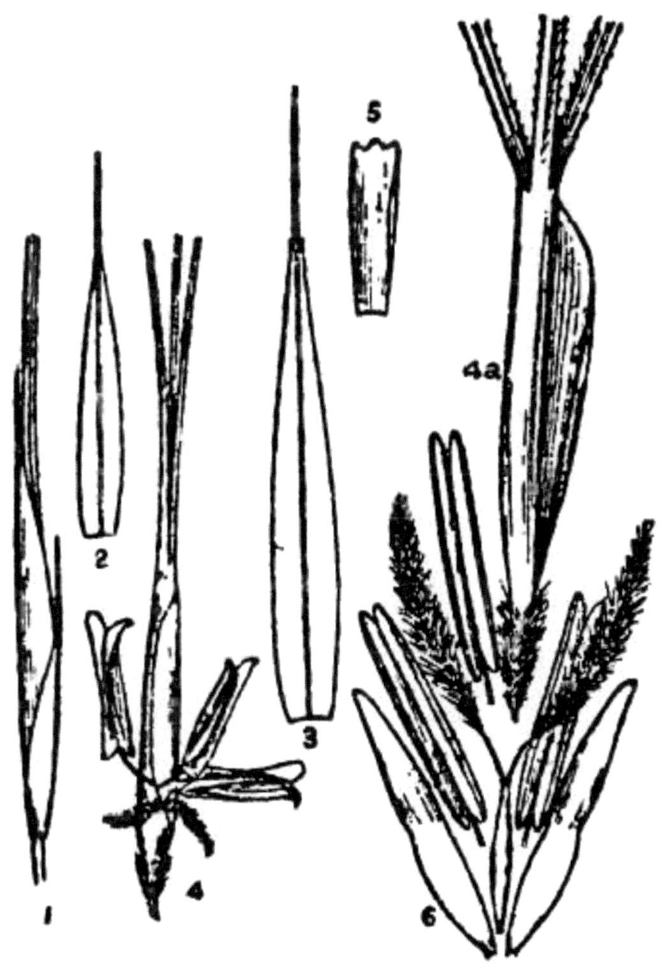

Fig. 174.—Aristida Hystrix.
1. A spikelet; 2, 3 and 4. the first, second and the third glume, respectively; 4a. the third glume and its awns; 5. palea of the third glume; 6. lodicules, anthers and the ovary.

The *inflorescence* is an effuse panicle, as long as broad, varying in length from 4 to 10 inches; the main rachis is stout, finely scabrid, with stiff slender, horizontally spreading or reclining branches that arise in pairs from the nodes, the branches have swollen bases at the nodes and they are covered by long hairs.

The *spikelets* are 3/8 inch long excluding the awn. There are three *glumes*. The *first glume* is chartaceous, lanceolate, acuminate and terminating in an awn, 1-nerved, 3/8 to 1/2 inch including the awn, with the keel very finely scabrous. The *second glume* is longer than the first, chartaceous, lanceolate, terminating in an awn, 1/2 to 3/4 inch long including the awn, with a smooth keel. The callus of the third glume is short, pointed and villous. The *third glume* is chartaceous finely scabrid 1/4 to 3/8 inch long excluding the awn, 3-nerved, 3-lobed at the apex and the lobes becoming awns; *awns* are 1 inch long, the middle one being a little longer. The outer margin of the glume is broader than the inner margin and is rounded at the apex at the base of the awn. There are three *stamens* and the anthers are pale or purplish. The style branches are purplish. The *lodicules* are 1/8 inch long obliquely lanceolate.

This grass is fairly common in all open dry situations throughout this Presidency.

Distribution.—Deccan Peninsula and Ceylon.

[Pg 227]

Fig. 175. — Aristida mutabilis.

[Pg 228]

Aristida mutabilis, *Trin. & Rup.*

This is a small tufted annual grass with simple or branched slender stems spreading at the base, and sometimes geniculately ascending and rooting at the lower nodes, 6 to 12 inches long. The *nodes* have dark purple rings when dry.

The *leaf-sheath* is glabrous, with membranous margins and long hairs at the mouth. The *ligule* is a row of short dense hairs.

The *leaf-blade* is slender, convolute, rigid, curved, and the tip ending in a sharp point, 1 to 3 inches long.

The *inflorescence* is a narrow panicle, cylindric, with short crowded branches, some of them remote lower down, peduncle is smooth, and rachis smooth or scaberulous; branches and pedicels are scaberulous.

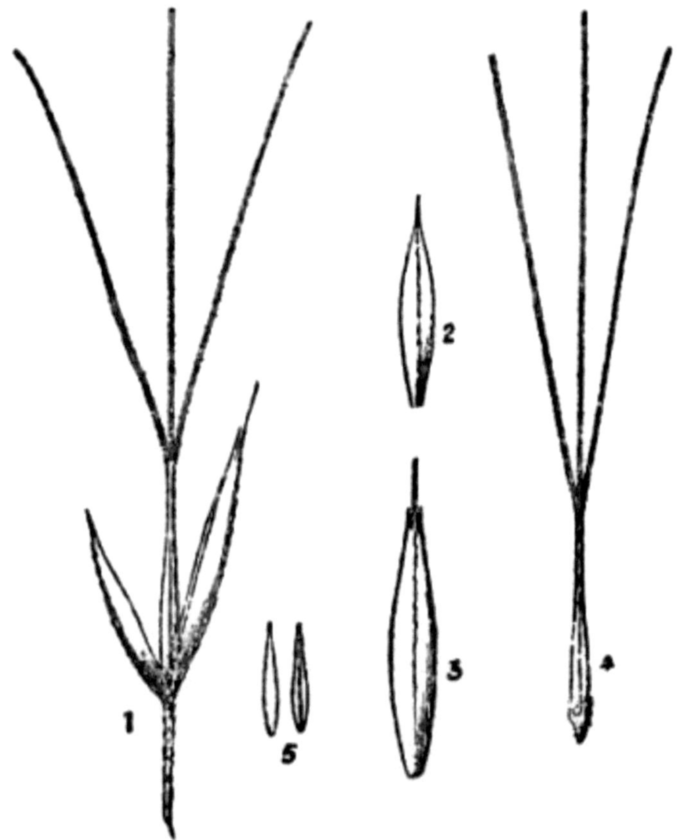

Fig. 176.—Aristida mutabilis.
1. A spikelet; 2, 3 and 4. the first, second and the third glume, respectively; 5. grain.

The *spikelets* are shortly pedicellate, pale-green about 1/4 inch long exclusive of the awn. There are three glumes. The *first glume* is membranous, oblong-lanceolate, shortly awned, 1-nerved, keeled and scaberulous on the keel and the sides. The *second glume* is narrower and longer than the first, shortly awned 1-nerved, 2-toothed, obscurely scaberulous and encircling the third glume. The *third glume* is narrow, convolute, scaberulous, 3-nerved awned with a shortly bearded callus, the awn is three branched articulate to the short column at the base about 3/4 inch long with the middle branch slightly longer than the other two; *palea* is minute. *Lodicules* are two and narrow. The grain is narrow as long as the glume and grooved.

This resembles in general habit and appearance *Aristida Adscensionis*, but it is not so widely distributed. So far this has been noticed only in Tinnevelly and Nellore districts.

Distribution.—Southern India, the Punjab and Rajputana, also in Arabia and tropical Africa.

[Pg 229]

Aristida funiculata, *Trin. & Rup.*

This is a slender annual grass with geniculately ascending stems, radiating on all sides. The stems vary in length from 10 to 20 inches.

The *leaf-sheath* is glabrous and cylindrical. The *ligule* is a short membrane ciliate at the margin, or a close set fringe of hairs.

The *leaf-blade* is flat or convolute, narrowly linear-acuminate, with long scattered hairs on the upper surface and tufts of long hairs at the mouth, and varying in length from 2 to 6 inches and in breadth from 1/20 to 1/12 inch.

The *inflorescence* is a narrow, lax panicle with short, erect, capillary branches. The spikelets vary in length from 1/2 to 7/8 inch.

There are three *glumes*. The *first glume* is linear-lanceolate, acute and terminating in an awn, 1-nerved and varying in length from

3/4 to 7/8 inch. The *second glume* is similar to the first, but narrower and shorter, 1/2 inch or longer. The *third glume* is very short, and is prolonged towards the apex as a narrow firmly convolute strap forming a twisted column of about an inch jointed at the base, and this ends in three slender scabrid awns of about 1-1/4 inch, the middle one being longer. The glume just below the joint is finely scabrid to a little distance. The *palea* is short. *Anthers* are small, purple. The *style* branches are also purple. *Lodicules* are oblong, obliquely truncate at the apex and about 1/10 inch long. The grain is cylindric.

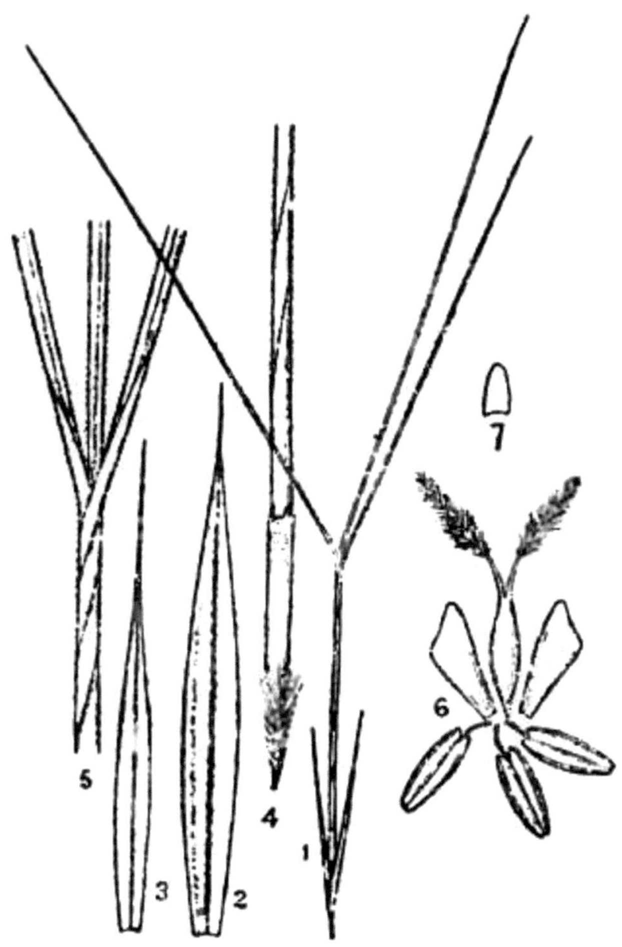

Fig. 177.—Aristida funiculata.
1. A spikelet; 2, 3 and 4. the first, second and the third glume, respectively; 5. a portion of the column at the top and the basal portions of the awns; 6. the ovary, lodicules and the stamens; 7. palea of
the third glume.

Found in open dry situation in several places, but not widely distributed.

Distribution.—From the Punjab to Concan and Madras Presidency, Arabia, Baluchistan and Tropical Africa.

[Pg 230]

32. Sporobolus, *Br.*

These are perennial or annual grasses with varied habit. Inflorescence is an open or contracted or spiciform panicle. Spikelets are small consisting of three membranous glumes, 1-nerved or nerveless. The first and the second glumes are unequal, persistent or separately caducous. The third glume is ovate or oblong, acute or obtuse, longer or shorter than the second, 1-nerved, paleate; palea is as long as the glume and of the same texture of the glume dorsally narrowly inflexed along the middle line and splitting into two halves. Lodicules are very minute or absent. Stamens one to three. Styles are with short stigmas. Grain oblong, obovoid or round.

KEY TO THE SPECIES.

- Glumes I and II both shorter than III.
    - Panicle rather narrow with short capillary branches; glumes I and II nerveless. 1. S. diander.
- Glume I shorter than II and III and II nearly or quite as long as III.
    - Panicle contracted, narrow and spiciform; glume I 1-nerved. 2. S. tremulus.
    - Panicle open and effuse.
        - Branches with spikelets and pedicels appressed. 3. S. coromandelianus.

        - Panicle short; leaves glabrous. Branches with pedicel and spikelets drooping and not appressed. 4. S. commutatus.
        - Panicle large; leaves with long hairs. 5. S. scabrifolius.

[Pg 231]

Sporobolus diander, *Beauv.*

This is a tufted annual or perennial grass. Stems are slender with leaves tufted at the base, 1 to 3 feet high.

The *leaf-sheath* is glabrous and smooth, ribbed, the lower short and the upper very long. *Nodes* are glabrous. The *ligule* consists of a fringe of minute hairs.

The *leaf-blades* are usually flat, glabrous, strongly nerved, with filiform tips, 3 to 10 inches by 1/25 to 1/16 inch.

The *inflorescence* is an erect narrow pyramidal panicle, varying in length from 4 to 10 inches and about 2 inches in breadth. The branches are very fine, spreading and in scattered fascicles, 1/2 to 2 inches long, with many very small spikelets arranged racemosely along the axis. *Spikelets* are small 1/18 to 1/20 inch long, with very short pedicels. The *first glume* is very short less than 1/5 inch, broadly oblong, nerveless, hyaline, broadly truncate and erose at the apex. The *second glume* is a little longer than the first, but shorter than the third, hyaline, broadly elliptic-oblong, nerveless or obscurely 1-nerved. The *third glume* is broadly ovate-oblong, subacute, 1-nerved, paleate; the *palea* is plicate in the median line. Stamens are usually two. The grain is obovoid, truncate at the apex, and with a small white swelling in the centre at the apex, rugulose, red-brown.

Fig. 178.—Sporobolus diander.
1. A portion of a branch; 2. a spikelet; 3, 4 and 5. the first, second and the third glume, respectively; 6. palea of the third glume; 7. anthers and the ovary.

This grass grows usually gregariously in somewhat sheltered situations all over the Presidency on the plains and low hills. This is an excellent fodder grass. It forms fairly large tufts with plenty of green leaves on rich moist soils. When the leaves are young cattle eat this grass very eagerly, but do not seem to care for it when the leaves become old. However by frequent grazing it can be made to produce young leaves in succession. This grass is also an excellent soil binder, as its roots form a perfect matting in any kind of moist soil soon after planting. This is very difficult to eradicate when once established.

Distribution.—Throughout India and Burma.

[Pg 232]

Fig. 179.—Sporobolus tremulus.

[Pg 233]

Sporobolus tremulus, *Kunth.*

A small tufted perennial grass.

The plant consists of prostrate stems and stolons, filiform and wiry. Stems vary in length from 2 to 18 inches, prostrate or erect, rooting at the lower nodes; flowering branches always ascending.

The *leaf-sheath* is glabrous, finely striate, shorter than the internode. The *ligule* is a very short ciliated membrane.

The *leaf-blade* is narrow linear, pungent, somewhat rigid, flat, distichous, base rounded with or without a few long hairs and varies in length from 1/4 to 1 inch and in breadth from 1/20 to 1/16 inch, but in plants growing in rich moist soils the leaves become longer reaching 3-1/2 inches in length.

The *inflorescence* is a narrow spiciform panicle with appressed branches and spikelets, sometimes interrupted, varying in length from 3/4 to 1-1/4 inch; both the peduncle and the main rachis are glabrous, and the latter wavy.

Fig. 180.—Sporobolus tremulus.
1. Spike; 2. spikelet; 3 and 4. first and second glumes; 5 and 6. third glume and its palea; 7. ovary and anthers.

The *spikelets* are 1/16 inch long, oblong-lanceolate, pale, crowded, glabrous, shortly pedicelled on thinly scaberulous filiform short branches. There are three glumes in the spikelet, and all the glumes are membranous and thin. The *first glume* is a little shorter than the second and about two-third the length of the third glume and 1-nerved. The *second glume* is a little shorter than the third or equal to but not longer, oblong-lanceolate, subacute or obtuse, 1-nerved and obscurely scaberulous at the back along the nerve. The *third glume* is broadly oblong, subacute or obtuse, 1-nerved, glabrous, with a palea as long as the glume; the *palea* is 2-nerved, oblong and truncate at the apex. *Stamens* are three and anthers are pale greenish yellow. *Stigmas* are pale. *Lodicules* are two, small.

This grass is an excellent one for binding the soil and may also prove successful as a fodder grass. It usually flourishes in moist situations, in sandy loams and rich heavy soils.

Distribution.—Plains throughout India and Ceylon.

[Pg 234]

314

Fig. 181.—Sporobolus coromandelianus.

[Pg 235]

Sporobolus coromandelianus, *L.*

The plant is a densely tufted annual varying in size with the nature of the soil, small and stunted in hard dry soils and large and spreading in rich loose and moist soils.

The stems are closely spreading on the ground, rooting sometimes at the lower nodes, branching freely, profusely leafy at the base, covered by a few scale leaves, and 2 to 12 inches long.

The *leaf-sheath* is glabrous, faintly and finely striate, distichously imbricate, compressed, somewhat keeled, outer margin ciliate, and bearded at the mouth. The *ligule* is a thin short membranous ridge with a fringe of dense fine hairs. The leaf-sheath enclosing the base of the peduncle is rather long, glabrous with a tuft of short hairs at the mouth.

The *leaf-blade* is green without any glaucousness about it, 1/2 to 6 inches long, 3/16 to 1/4 inch broad, lanceolate or linear-lanceolate, flat, acuminate, slightly coriaceous, many-nerved with a prominent midrib, scaberulous throughout, with a few long scattered deciduous, tubercle-based hairs towards the base, base subcordate, margin cartilaginous, scabrid and finely serrulate.

Fig. 182.—Sporobolus coromandelianus.

1. Portion of a spike showing the verticillate arrangement of the branches and the glands; 2. spikelet; 3. first glume; 4 and 5. second and third glumes; 6. palea of the third glume; 7. anthers and ovary; 8. grain.

The *inflorescence* is a pyramidal panicle 1-1/2 to 4 inches long, erect on a terete glabrous peduncle 1-1/2 to 6 inches long, the main rachis is slender, erect, striate, glabrous and has glandular streaks just above the insertion of the branches of the lowest verticil. Branches are capillary, stiff and spreading, horizontally verticillate or subverticillate, the lowest whorl consisting of five to sixteen or seventeen branches and the others from three to nine, shining, swollen at the point of insertion and provided with a glandular scar a little above the point of insertion; branchlets are very close, appressed to the rachis of the branch never drooping or spreading, each bearing two to five spikelets.

[Pg 236]

The *spikelets* are small, 1/20 to 1/16 inch subsessile or pedicelled, always appressed to the rachis solitary in the upper portions of the branches, and two to five on the branchlets in the lower portion, pale, green or rarely copper coloured, oblong or lanceolate, acute or acuminate, caducous or glumes one and two persistent.

There are three *glumes*. The *first glume* is very small, hyaline, ovate, obtuse, occasionally truncate or acute, about one-fifth of the third glume or less. The *second glume* is membranous, ovate or oblong-lanceolate, acute or acuminate, thinly scaberulous and 1-nerved. The *third glume* is as long as or a little shorter than the second glume, 1-nerved and paleate. The *palea* is as long as the glume, oblong, 2-nerved, splitting in two portions between the nerves as soon as the grain is formed. *Stamens* are three with reddish purple anthers; *stigmas* are white at first, but turning brown while withering. *Lodicules* are two, minute. The grain is oblong, pale, brown and obtuse at both ends, embryo about 1/3 of the grain.

This grass flourishes in all kinds of soils all over the Presidency.

Distribution.—Throughout the plains of India and Ceylon. Also in Afghanistan and South Africa.

[Pg 237]

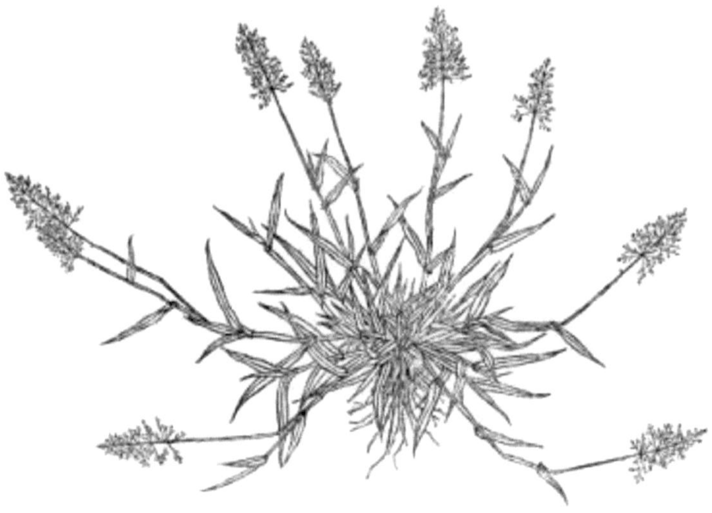

Fig. 183.—Sporobolus commutatus.

[Pg 238]

Sporobolus commutatus, *Kunth.*

This is an annual and usually grows in loose tufts. Stems are slender, always erect or ascending, leafy and branching, 2 to 15 inches long.

The *leaf-sheath* is shorter than the internode, slightly compressed, finely striate, glabrous and occasionally with a few scattered tubercle-based hairs, margin ciliate; the uppermost sheath is cylindric somewhat long and embraces the greater portion of the peduncle and has a bunch of short hairs at the top.

The *leaf-blade* is narrow linear-lanceolate, acuminate scaberulous throughout, with long tubercle-based hairs scattered all over, but more of them near the base; margins spinulosely distantly serrulate or scabrid, base rounded or subcordate, 1/2 to 4-1/2 inches long and 1/16 to 3/16 inch wide.

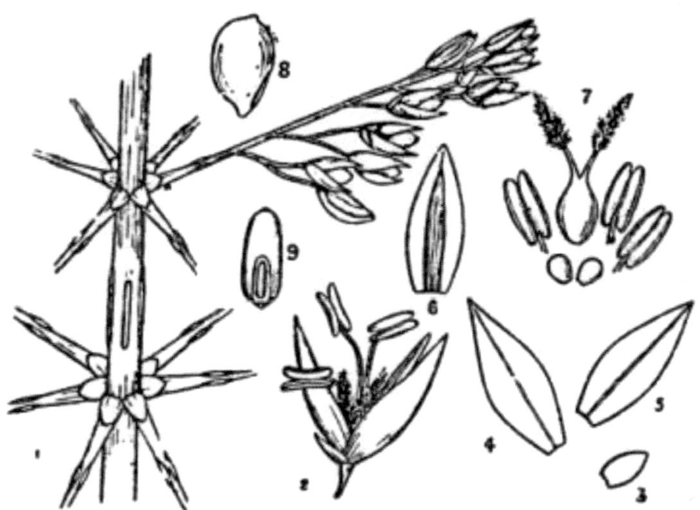

Fig. 184.—Sporobolus commutatus.
1. A portion of a branch; 2. spikelet; 3, 4 and 5. first, second and the third glume; 6. palea of the third glume; 7. ovary and anthers; 8 and 9. grain.

The *inflorescence* is diffuse, pyramidal, 1 to 3 inches by 3/4 to 2 inches, on a slender glabrous peduncle 1 to 6 inches long, main rachis is slender and angled, with a glandular streak or without it. Branches are effuse, fine, capillary (more so than in S. coromandelianus), obliquely ascending, never stiff and horizontal, verticillate or irregularly subverticillate, the lowest whorl of five to twelve and the others three to seven branches; the rachis of the branches is obscurely scaberulous, slightly swollen at the point of insertion; branchlets are never appressed to the branch, always drooping and spreading on all sides, and bearing two to four spikelets.

The *spikelets* are about 1/16 inch long, ovate-lanceolate, acute or acuminate dark or pale green, sometimes purplish, solitary or two to four on long slender pedicels, drooping, never appressed, and with glandular streaks. There are three *glumes*. The *first glume* is minute, hyaline, ovate, obtuse or acute, nerveless. The *second glume* is five or six times as long as the first, ovate lanceolate, 1-nerved, acuminate. The *third glume* is equal to or a [Pg 239] little shorter than the second, ovate-lanceolate, acute, 1-nerved paleate; *palea* is equal to the third glume, 2-nerved splitting into two halves between the nerves. *Anthers* are three and purple in colour. *Stigmas* are white and feathery. Grain as in *S. coromandelianus.*

In Flora of British India, this plant is included under Sporobolus coromandelianus. These two plants (*S. coromandelianus* and *S. commutatus*) are quite distinct and grow side by side. As the differences are not easily seen in herbarium specimens the two plants are put together under the one species *S. coromandelianus.* The branches are tufted and are usually decumbent at base, leaves quite green and somewhat broad in *S. coromandelianus;* and in *S. commutatus,* branches are usually not decumbent at base, generally erect from the base and leaves are green glaucous and somewhat narrow. The most striking difference, however, is in the panicle. The branches of the panicle are always stiff and horizontal in S. coromandelianus and the spikelets are appressed to the branches and never spreading or drooping, whereas in *S. commutatus* the branches are never stiff and horizontal, always obliquely ascending and the spikelets are spreading and drooping. Judging from living plants these two are undoubtedly distinct and so this plant is treated as a distinct species retaining Kunth's name *Sporobolus commutatus.* Enumeratio Plantarum, Pl. I, 214.

Distribution.—This occurs in Coimbatore, Madras and Bellary Districts; but it is not so common nor so widely distributed as S. coromandelianus, *L.*

[Pg 240]

Fig. 185.—Sporobolus scabrifolius.

[Pg 241]

Sporobolus scabrifolius, *Bhide.*

The plant is a very pretty one, especially when in flower. It is a loosely tufted annual varying in height from 5 to 30 inches. Stems are slender, terete, 6 to 30 inches long, bent at the base, then geniculately ascending and finally becoming erect, glabrous, pale green or purplish.

The *leaf-sheath* is shorter than the internode, slightly compressed, obscurely keeled, glabrous and striate, margin is thinly ciliate on one side, especially towards the mouth which is bearded. The leaf-sheath embracing the peduncle is longer than the lower sheaths. The *ligule* is a fringe of close-set hairs on an inconspicuous ridge. The *nodes* are glabrous.

The *leaf-blade* is glaucous green, 1 to 5 inches long, 1/8 to 3/8 inch broad, linear-lanceolate or lanceolate, acuminate, flat, rounded or subcordate, and amplexicaul at base, scaberulous throughout, with tubercle-based deciduous hairs on both the surfaces, and bearded at the base above the ligule; the margin is thickened, serrulate, ciliate with bulbous-based deciduous hairs.

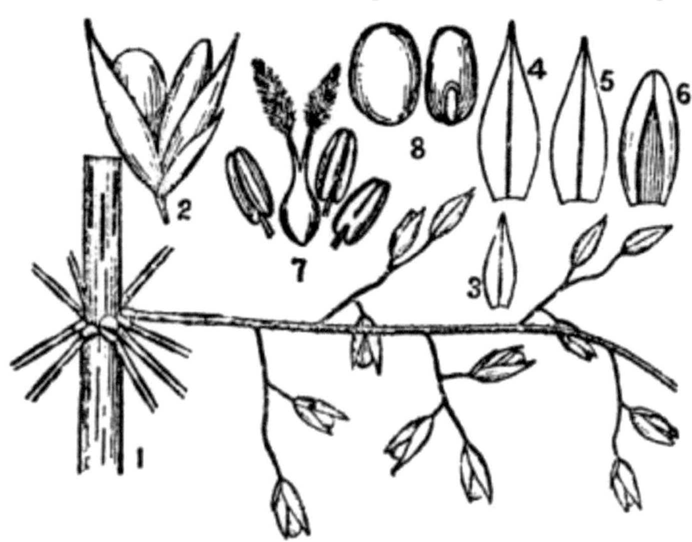

Fig. 186.—Sporobolus scabrifolius.
1. Portion of a branch; 2. spikelet; 3, 4 and 5. the first, second and third glumes; 6. palea; 7. anthers and ovary; 8. grain.

The *inflorescence* is an effuse panicle, 2-1/2 to 7 inches long and 1 to 4-1/2 inches broad, pyramidal or elliptic on a slender peduncle 1 to 7 inches long; *rachis* is striolate, cylindric, glabrous and partly green and partly purplish. Branches are capillary, 1/2 to 2-1/2 inches long, those in the middle of the panicle are often the longest pale green at first but turning purple later, whorled regularly or irregularly, with often a solitary or twin branches intervening, spreading, horizontal, reflexed, rarely one or two erect, dividing into still finer branchlets below, ending in a few solitary spikelets above, swollen at the base near the place of insertion and naked to a short length, scabrid. The lowest whorl consists of five to ten branches and in others they vary from three to eight; the branchlets are spreading and drooping bearing from two to seven spikelets. There are glandular streaks at the base of the branches above the point of insertion in the naked portion and also on the pedicels of the spikelets.

[Pg 242]

The spikelets are 1/20 to 1/16 inch long, lanceolate, acuminate, on finely capillary pedicels long or short, pale at first and becoming purplish when old. There are three *glumes*, the first two being empty. All the glumes are 1-nerved and membranous. The *first glume* is membranous, about two-thirds of the second, sometimes less, ovate-lanceolate, acuminate slightly scaberulous on the keel. The *second glume* is a little longer than the third, ovate-lanceolate, acuminate, scaberulous on the keel. The *third glume* is oblong-ovate, glabrous, flower bearing, paleate; the *palea* is shorter than the glume, 2-nerved, splitting into two between the nerves. *Anthers* are three, small, pale yellow at first but becoming purple when old, *stigmas* are pale. *Lodicules* are two and minute. Grain is rounded, slightly compressed, oblique at the base, nearly as long as broad.

Distribution. — In black cotton soils in Coimbatore and Bellary districts.

[Pg 243]

33. Gracilea, *Koen.*

These are small tufted grasses. The inflorescence is a spike bearing unilaterally turbinate clusters of spikelets which are 2-flowered.

The spikelets have usually four, and rarely six glumes and very often the rachilla is produced beyond the fourth glume. The first and the second glumes are narrow (the first being the narrowest), rigid, ciliate with long hairs and awned. The third glume is bisexual, chartaceous, broadly ovate, 3-nerved, shortly awned. The fourth glume is similar to the third but smaller and male. The fifth and sixth glumes when present are small and empty. Lodicules are two and small. Grain linear oblong.

KEY TO THE SPECIES.

- Stems stout; leaves not filiform; tip of glume III entire 1. G. nutans.
- Stems slender, leaves filiform; tip of glume III toothed 2. G. Royleana.

[Pg 244]

Gracilea nutans, *Koen.*

This grass is a perennial with stout fibrous roots. Stems are stout, leafy and creeping below, ascending later; naked and slender above, 4 to 10 inches long.

The *leaf-sheath* is glabrous, shorter than the blade, coriaceous and open above. The *ligule* is a ridge of hairs.

The *leaf-blade* is lanceolate, narrowed from the rounded or sub-cordate base to the acute tip, coriaceous, 3/4 to 1 inch long; margins are ciliate with tubercle-based cilia; the surfaces with or without a few scattered long tubercle-based hairs.

The *inflorescence* is 1 to 3 inches long, consisting of distant sessile fascicles of four to six spikelets; the *rachis* of the spike is flexuous; the *rachis* of the fascicles ends in three subulate empty glumes.

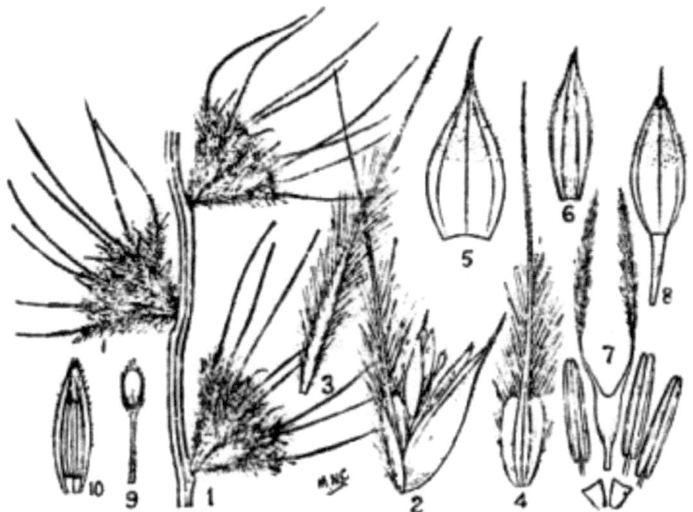

Fig. 187.—Gracilea nutans.
1. A portion of the inflorescence with three fascicles of spikelets; 2. a spikelet without the first glume; 3, 4, 5 and 8. the first, second, third and the fourth glume, respectively; 6 and 10. palea of the third and the fourth glume, respectively; 7. lodicules, stamens and the ovary; 9. the rachilla produced beyond the fourth glume.

The *spikelets* are closely appressed and each one has four *glumes*. The *first* and the *second glumes* are empty, 2/5 inch long, rigidly coriaceous, gradually narrowed from a villous base into an erect, scabrid awn, 1-nerved. The *second glume* has broad hyaline margins towards the base. The *third glume* is about 1/10 inch, ovate, with a short scabrid awn at the tip, scaberulous at the back just above the middle, 3-nerved, paleate and with both stamens and ovary; *palea* is narrow, lanceolate, as long as the glume and 2-toothed at the tip. The grain is oblong, brownish. The *fourth glume* is about half as long as the third glume, with a short, stout, smooth rachilla, ovate-lanceolate, terminated at the tip by two teeth and a short awn, scabrid above the middle at the back, paleate and male; *palea* is shorter than the glume; the rachilla is produced beyond the fourth glume and terminates in a thickening.

325

This grass grows in open somewhat dry loamy and laterite soils in the East Coast districts.

Distribution. — Mysore and the Carnatic and Ceylon.

[Pg 245]

Gracilea Royleana, *Hook. f.*

This is a slender annual grass. Stems are very slender, densely tufted, geniculately ascending or erect, 3 to 8 inches long.

The *leaf-sheath* is either covered with scattered tubercle-based hairs or glabrous. The *ligule* is a hairy ridge. The *nodes* are glabrous.

The *leaf-blade* is filiform, linear-lanceolate, acutely pointed, glabrous or nearly so, margins distantly ciliate, 1 to 2 inches long by 1/16 inch or less.

The *inflorescence* is 1/2 to 3 inches long and consists of fascicles of spikelets; the rachis is trigonous, smooth, and flexuous.

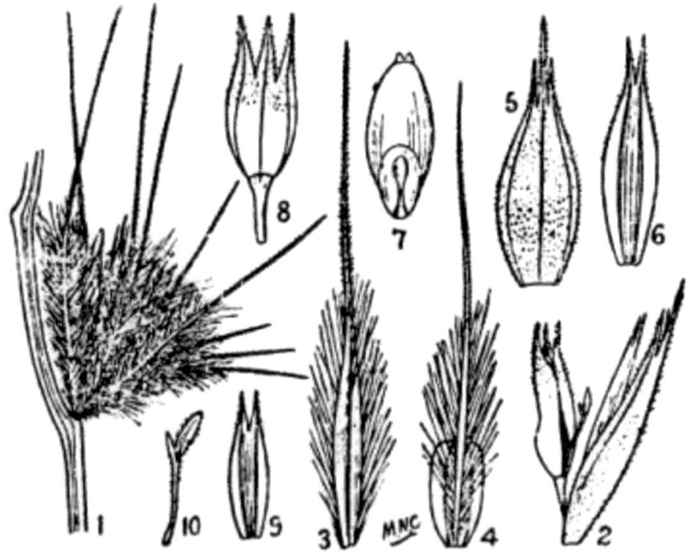

Fig. 188. — Gracilea Royleana.
1. A fascicle of spikelets; 2. the spikelet without the first and the

326

second glumes; 3,4, 5 and 8. the first, second, third and the fourth glume, respectively; 6. palea of third glume; 7. grain; 9. palea of the fourth glume; 10. rachilla.

The *spikelets* consist of four *glumes*. The *first glume* is rigidly coriaceous, gradually narrowed from a villous base to an erect scabrid awn, 1-nerved. The *second glume* is also coriaceous, narrowed to an awn but has broad hyaline margins towards the base. The *third glume* is ovate-lanceolate, scabrid all over the back and with two teeth, one on each side of the awn, paleate; the *palea* is 2-toothed at the apex and as long as the glume and contains three stamens and the ovary. The grain is oblong brownish. The *fourth glume* is stalked, shorter than the third glume, distinctly 3-toothed at the apex, scabrid at the back above the middle, paleate and male; the *palea* is smaller than the glume and 2-toothed at the apex. The *rachilla* is produced behind the palea and it ends in two small teeth, one being slightly larger than the other.

This grass is a very slender one and it is closely allied to *Gracilea nutans*. It differs from *G. nutans* in being an annual and in having filiform leaves, bicuspidate third glume which is scabrid all over the back and a fourth glume distinctly tricuspidate at the apex. This does not occur so widely as *Gracilea nutans*.

Distribution.—Bellary and Chingleput districts, the Punjab, Rajputana, Concan and Kanara.

[Pg 246]

34. Enteropogon, *Nees.*

Tall slender grasses with very long narrow leaves. Spikelets are 2-flowered, narrow, biseriate, unilateral, imbricate on the rachis of a solitary spike; the rachilla is elongate between the flowering glumes and produced beyond them and terminates in a rudimentary awned glume. There are four glumes. The first two glumes are hyaline, unequal-nerved and persistent. The third and the fourth glumes are chartaceous, narrowly lanceolate, 3-nerved, bicuspidate and awned below the tip; awns are capillary, straight; the callus is bearded and articulate at the base. The third glume bears a bisexual or female flower and the fourth bisexual or male. Lodicules are two. Stamens are three with long anthers. Styles short diverging from the base, with short stigmas laterally exserted.

[Pg 247]

Enteropogon melicoides, *Nees.*

This is a tall perennial grass with stout roots. Stems are densely tufted on a short woody root-stock, erect, leafy, 1 to 3 feet long.

Leaf-sheaths are compressed and distichous below, glabrous or sometimes with a few hairs close to the margin. Ligule is a ridge with long hairs.

The *leaf-blade* is very long 1/6 to 1/4 inch broad, auricled at the base, narrowed into very finely acuminate or capillary tips midrib prominent; scaberulous on both the surfaces and with long hairs on the auricles.

The *spikes* usually solitary, but occasionally binate, 6 to 10 inches long; rachis is quite smooth and dorsally rounded.

Fig. 189.—Enteropogon melicoides.
1. A portion of the spike; 2 and 3. the first and the second glumes; 4. the spikelet with its callus, flowering glumes and the rachilla; 5 and

8. the third and the fourth glume; 7. the fourth glume and the rachilla; 6 and 9. palea of the third and the fourth glume; 10. ovary, stamens and lodicules; 11. grain front and back view.

The *spikelets* are about 1/4 inch long, erecto-patent. There are four *glumes*. The *first glume* is lanceolate, 1-nerved, and persistent. The *second glume* is twice as long as the first, linear-lanceolate, with a very short awn and 2-toothed at the tip, 1-nerved, persistent. The *third glume* is rigid, lanceolate-linear, 3-nerved, scaberulous all over; paleate and awned; awn is nearly as long as the glume, rigid. The *fourth glume* is similar to the third glume in all respects but shorter. The rachilla is produced beyond the fourth glume and it terminates in an awned rudimentary glume. The third glume as well as the fourth glume contains a perfect flower and the grain is developed always in the third and mostly in the fourth also. The grain is oblong, brownish, dorsally concave and ventrally raised and convex. The grain in the fourth glume is usually much smaller than that found in the third glume.

This usually grows amidst thickets and occurs all over this Presidency.

Distribution. — Mysore, Burma, Ceylon and Seychelle Islands.

[Pg 248]

35. Cynodon, *Pers.*

These are perennial grasses with stems creeping and rooting at the nodes, and producing tufts of barren branches and flowering stems at the nodes. The inflorescence consists of two to six spikes in terminal umbels. The spikelets are small, 1-flowered, laterally compressed, sessile, alternately 2-seriate and imbricate on one side of the rachis. The spikelet has three glumes. The first two glumes are empty, thin, keeled, and acute or mucronate. The third glume is the largest, boat-shaped, 3-nerved, with ciliate keels, palea is 2-keeled, somewhat shorter than the glume. Lodicules are two. The anthers are somewhat large. Grain is oblong, free.

KEY TO THE SPECIES.

- Glumes I and II shorter than III.

- o   Underground stems present.
  - ▪ Hairs on the margins and keels of glume III pointed and not clavate. 1. C. dactylon.
- o   Underground stems absent.
  - ▪ Hairs on the margins and keels of glume III clavellate and pointed at the apex. 2. C. intermedius.
- Glume I shorter than II but II equal to or longer than III —
  - o   Hairs on the margins and keels of glume III clavellate and rounded at the apex. Underground stems absent. 3. C. Barberi.

[Pg 249]

Fig. 190.—Cynodon dactylon.

[Pg 250]

Cynodon dactylon, *Pers.*

This is a perennial grass with creeping branches and also with numerous deeply penetrating underground stems covered with white scale-leaves. Stems are prostrate, widely creeping and rooting at the nodes and forming matted tufts with slender, erect or ascending flowering branches, 3 to 12 inches high.

The *leaf-sheath* is somewhat tight, glabrous, membranous at the mouth which is villous. The *ligule* is a fine ciliate rim.

The *leaf-blade* is soft, narrowly linear, finely acute, acuminate or pungent, somewhat glaucous, conspicuously distichous at the base of the stem and, in non-flowering branches, scabrid along the margins.

The *inflorescence* consists of two to eight smooth, digitate, green or purplish spikes, 1 to 3 inches long; *rachis* is slender, compressed or angular, scaberulous.

Fig. 191. — Cynodon dactylon.
1. A portion of spike, front view; 2. back view of a bit of spike; 3. spikelet; 4. first glume; 5. second glume; 6. third glume; 7. palea of third glume and rachilla; 8. lodicules, ovary and anthers; 9. hairs on the margin and keel of third glume.

Spikelets are laterally compressed, sessile, imbricate, arranged alternately in two series along one side of the rachis; *rachilla* produced beyond the first two glumes and hidden at the back of the palea between the two keels, small, slender and blunt when old and with a membranous imperfect glume when young, less than half the length of the spikelet. There are three *glumes*. The *first* and *second glumes* are shorter than the third, empty, ovate-lanceolate, acute, membranous with one thick green nerve in the middle, keeled, upper margin and keel scaberulous. The *second glume* is usually a little longer than the first, but occasionally also slightly shorter than the first. The *third glume* is longer than both the first and second glumes, obliquely oblong to ovate, subacute, membranous, boat-shaped, smooth, keeled, 3-nerved, one central along the keel and two marginal, keel scabrid below with stiff pointed hairs above, tip and lower margins scabrid or pilose, *palea* linear oblong, a little less than the third glume, obtuse, 2-nerved and with two scabrid keels. *Stamens* are three with pale purple anthers. *Lodicules* are two. Stigmas are purplish. Grain is oblong, slightly flattened, dorsally rounded, dull reddish-brown.

This is the common Hariali grass. It is also called "Devil's grass."

Distribution. — It is cosmopolitan.

[Pg 251]

Fig. 192. — Cynodon intermedius.

[Pg 252]

Cynodon intermedius, *Rang. & Tad.*

This grass is a widely creeping perennial.

The stems are slender, glabrous, creeping superficially and root-ing at the nodes, but never rhizomiferous, leafy with slender erect or geniculately ascending flowering branches, and varying in length from 12 to 18 inches. *Nodes* are slightly swollen, glabrous, green or purplish.

The *leaf-sheath* is smooth, glabrous, slightly compressed, sparsely bearded at the mouth, shorter than the internode, except the one enclosing the peduncle which is usually long. The *ligule* is a shortly ciliated rim.

The *leaf-blade* is linear, flat, finely acuminate, scaberulous above and along the margins, smooth below except in some portions of the midrib, 1/2 to 7 inches in length and 3/16 to 1/4 inch in breadth.

Fig. 193.—Cynodon intermedius.
1 and 2. Front and back view of a portion of a spike; 3. a spikelet; 4. first glume; 5. second glume; 6. third glume; 7. palea with the rachil-

la at its back; 8. lodicules, stamens and the ovary; 9. clavellate and pointed hairs of the margins and keel of the third glume (very much enlarged); 10. grain.

The *inflorescence* consists of four to eight long, thin, slender, slightly drooping, digitately arranged spikes, 2 to 4 inches long on a long smooth peduncle; the rachis is tumid and pubescent at its base, slender, somewhat compressed and scaberulous.

The *spikelets* are rather small, narrow, greenish or purplish, 1/15 inch long or less, the rachilla is slender, produced to about half the length of the spikelet behind the palea. There are three *glumes*. The *first* and the *second glumes* are lanceolate acute or acuminate, 1-nerved, keeled, keel obscurely scabrid, very unequal, the first glume being always shorter than the second glume. The *third glume* is obliquely ovate-oblong, chartaceous, longer than the second glume, obtuse or subacute and 3-nerved; the margins and keel with close set clavellate hairs pointed at the apex; *palea* is chartaceous, 2-keeled, keels obscurely scaberulous and without hairs. There are three *stamens* with somewhat small purple anthers. *Ovary* with purple stigmas and two small *lodicules*. Grain is oblong reddish brown, with a faint dorsal groove.

[Pg 253]

This species is closely allied to the cosmopolitan species *Cynodon dactylon*, Pers. and to another new species *Cynodon Barberi*, Rang. & Tad. described in the "Journal of the Bombay Natural History Society," Volume 24, part IV, page 846, and it is therefore named *Cynodon intermedius*. (See Journal of the Bombay Natural History Society, Volume 26, part I, pages 304 and 305.) This grass differs from *Cynodon dactylon*, Pers. (1) in not having underground stems and having only stems creeping and rooting along the surface of the ground, (2) in having less rigid leaves, (3) by having longer, slenderer, somewhat drooping spikes and narrower spikelets, (4) by having the first two glumes always unequal, the second being longer, (5) by having clavellate pointed hairs on the margins and keels of the third glume and (6) by having smaller anthers. Compared with *Cynodon Barberi*, this plant is more extensively creeping with longer slender branches and the leaves are usually very much longer, and the third glume is longer than the second.

Distribution.—So far, this was collected at Gokavaram in Gōdāvari district No. 8262, in Chingleput No. 11488, in Tinnevelly district Nos. 13129 and 13259, and at Kallar on the Nilgiris No. 13988.

[Pg 254]

Fig. 194.—Cynodon Barberi.

[Pg 255]

Cynodon Barberi, *Rang. & Tad.*

This grass is perennial with slender, creeping stems, 12 to 24 inches long, rooting at the nodes and invariably with two or three rarely more branches from each node; flowering branches are slender, erect or ascending, 1 to 6 inches long.

The *leaf-sheath* is short, smooth, compressed with scattered long hairs at the mouth. The *ligule* is a narrow membrane with the edge cut into narrow lobes.

The *leaf-blade* is flat, linear, acute or subacute, scaberulous, 1/3 to 3-1/2 inches long, 1/8 to 3/16 inch broad.

336

Fig. 195.—Cynodon Barberi.
1. Front and back view of a portion of spike; 2. a single spikelet; 3. a
spikelet with the flower out; 4. the third glume, its palea and the
produced rachilla with a minute glume; 5. clavellate hairs; 6. ovary;
7. lodicules; 8. grain.

The *inflorescence* consists of three to five digitate spikes, 3/4 to 1-
1/2 inches long, erect or spreading, pale green or purplish. The
spikelets are compressed laterally, sessile or obscurely pedicelled,
imbricate, alternately biseriate on the ventral side of the rachis, 1-
flowered; the *rachilla* is produced into a bristle behind the palea,
with or without a minute glume. There are three *glumes*. The *first
glume* is lanceolate, acute, shorter than the second, with a keel which
is scabrid. The *second glume* is lanceolate, acuminate, equal to or a
little longer than the third glume with a scabrid keel. The *third glume*
is obliquely oblong to ovate, subacute, truncate or 2-toothed, boat-
shaped, sub-chartaceous, 3-nerved, paleate and distinctly keeled;
the keel and the margins of the glume are densely covered with
distinctly clavellate hairs; *palea* is firmly membranous, equal to or
slightly smaller than the glume, linear-oblong, 2-keeled, densely
hairy with clavellate hairs along the keels, and 2-nerved. There are
two *lodicules* and three *stamens*. The *ovary* is ovoid with two style

branches. Grain is free within the glume, oblong, smooth, transparent, and the embryo is about one-third the length of the grain.

[Pg 256]

This species is closely allied to *Cynodon dactylon*, Pers., but differs from it in the following respects:—The absence of stoloniferous underground branches, leaves short and not finely pointed; spikes not exceeding five; the *second glume* is always equal to or longer than the *third glume*; presence of clavellate hairs on the keels and margins of the third glume and on the keels of the palea.

Distribution.—So far collected in Coimbatore, Salem, Tinnevelly, Chingleput and Gōdāvari districts.

[Pg 257]

36. Chloris, *Sw.*

These are annual or perennial grasses. Spikes are solitary or many in terminal umbels or short racemes, erect or spreading. Spikelets are unilateral, sessile, crowded, biseriate on a slender rachis with four to six glumes and 1 to 3-flowered; the rachilla is produced and disarticulating above the empty glumes. The first two glumes are unequal, narrow, keeled, membranous, 1-nerved, persistent, acute, mucronate and the second glume awned shortly. Floral glumes narrow or broad, acute, obtuse or minutely 2-toothed and awned, paleate; sterile glumes are small, without palea. There are two lodicules and anthers are rather small. Grain is narrow and free.

KEY TO THE SPECIES.

- Spikelets 1-flowered.
  - Perennial.
    - Rachilla produced beyond the flowering glumes and bearing awns with rudimentary glumes.
      - Spikes 4 to 10, long, whorled; spikelets narrow fusiform; glume III oblong lanceolate. 1. C. incompleta.

338

- Rachilla produced beyond the flowering glume and bearing 1 to 3 reduced glumes.
  - Spikes free at the base, digitate.
    - Spikes 6 to 9; spikelets 2-awned; glume III ovate, bearded with long hairs above the middle. 3. C. virgata.
    - Spikes 4 to 20; spikelets 3-awned; glume III broadly ovate, densely bearded dorsally and on the margins above the middle. 4. C. barbata.
  - Spikes connate at the base, erect and not spreading.
    - Spikes 2 to 6; spikelets narrow 4-awned, glume III ovate-lanceolate, bearded only on the margins and not at the back. 6. C. montana.
  - Annual.
    - Spike solitary, spikelets broadly cuneiform, 3-awned, glume III broadly cuneate, upper margins naked and keel villous. 2. C. tenella.
- Spikelets 1- to 3-flowered.
  - Perennial.
    - Spikes 5-9, spikelets broadly cuneate 3 to 5-awned, glume III bearded all through the margin and dorsally. 5. C. Bournei.

[Pg 258]

Chloris incompleta, *Roth.*

This is a perennial grass. Stems are procumbent when growing in open places, but erect if growing amidst bushes, often branched, ending in long naked peduncles, varying in length from 1-1/2 to 4 feet. In some cases prostrate stems produce roots at the nodes.

The *leaf-sheaths* are long, glabrous, the mouth being generally hairy. The *ligule* consists of long hairs. *Nodes* are glabrous.

The *leaf-blades* are linear, flat, finely acuminate and narrowed into very long points at the apex; glabrous or slightly hairy at the base and contracted, 4 to 10 inches long and 1/6 to 1/4 inch broad.

The *inflorescence* consists of two to five rarely six, very slender spikes, 3 to 8 inches long, forming a terminal whorl. The rachis is fine and scabrid.

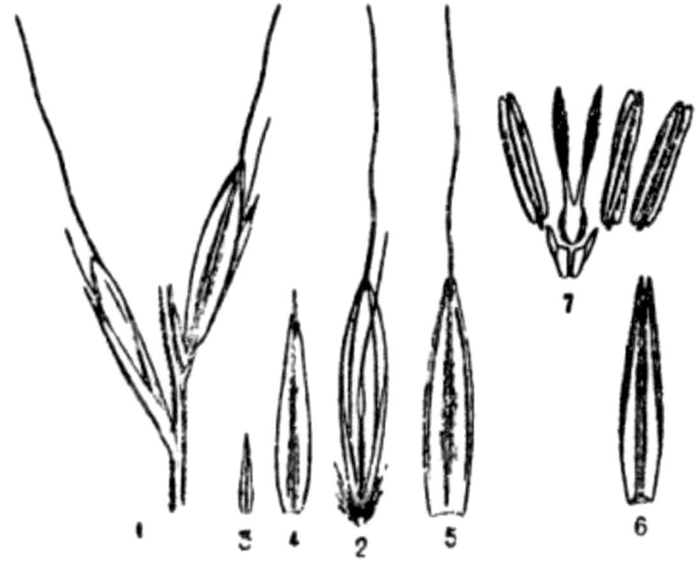

Fig. 196.—Chloris incompleta.
1. A portion of the rachis with two spikelets; 2. the third glume and its palea with the rudimentary fourth glume; 3 and 4. the first and the second glumes; 5 and 6. the third glume and its palea; 7. the ovary, anthers and lodicules.

Spikelets are narrowly lanceolate, closely appressed and imbricate, 1/6 inch long excluding the awn and very variable. There are four *glumes* in the spikelet. The *first glume* is very small linear-lanceolate, acute, about 1/10 inch or less. The *second glume* is lanceolate, membranous, three times the length of the first glume, 2-toothed at the apex and the mid-nerve produced into a very short awn. The *third glume* is oblong-lanceolate as long as the second glume or longer, 2-toothed at the apex, awned, the awn being about 3/8 inch long; the callus is bearded at the base. The palea is as long as the glume, 2-toothed or not at the apex. The *fourth glume* is very minute, awned and is borne by a rachilla produced to half the length of the third glume.

This grass is fairly common and grows in all situations and in all sorts of soils.

Distribution. — This occurs all over the Presidency in the plains.

[Pg 259]

Chloris tenella, *Roxb.*

This grass is a very slender annual with weak stems, branched from the base, 10 to 18 inches long.

The *leaf-sheath* is glabrous, compressed and keeled. The *ligule* is a truncate membrane. The *nodes* are glabrous.

The *leaf-blade* is linear to linear-lanceolate, flaccid, finely acuminate with the margin more or less ciliate towards the base, 3 to 8 inches long and 1/8 to 1/4 inch wide.

The *spikes* are solitary, erect. 1 to 2-1/2 inches long.

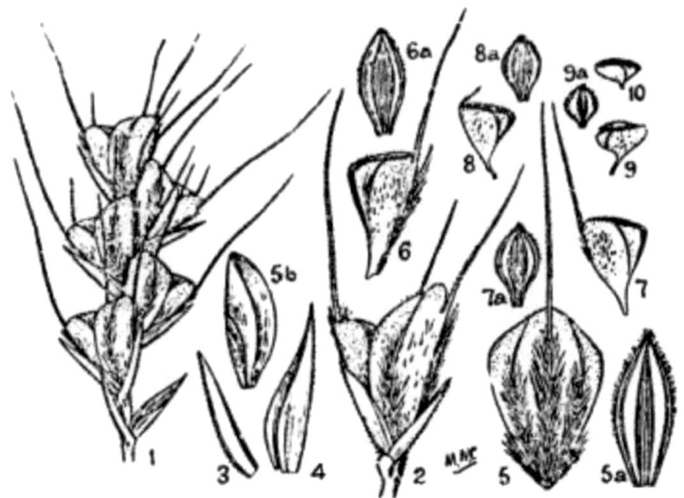

Fig. 197.—Chloris tenella.
1. A portion of the spike; 2. a spikelet; 3, 4, 5, 6, 7, 8, 9 and 10. the glumes in regular order beginning with the first; 5a, 6a, 7a, 8a and 9b. are the palea of the third, fourth, fifth, sixth and the seventh glumes, respectively; 5b. grain.

The *spikelets* are large about 1/4 inch long cuneate and bifarious. There are usually five to six *glumes* (and rarely up to eight). The *first glume* is ovate-lanceolate, acute and hyaline, 1-nerved. The *second glume* is a little longer and broader than the first glume, 1-nerved and this mid-nerve produced into a very short awn. The *third glume* is as long as the second or longer, coriaceous, obovate and truncate at the top, 3-nerved and the marginal nerves distant from the margin, keel and the lateral nerves villous to about three-fourths their length, scabrid at the apex close to the truncate margin, paleate; *palea* is elliptic, with ciliate margins, callus is densely villous. The *fourth glume* is nearly half or a little more than half of the third glume, narrower, paleate; *palea* is elliptic. The succeeding glumes *fifth* to the *eighth* are similar to the fourth in shape but they get smaller and smaller and the last glume is epaleate. The third glume is usually grain bearing, but rarely the fourth also may contain a grain, the remaining glumes being sterile. Grain is oblong, lenticular, brownish.

This grass is widely spread in the Ceded districts and appears to be a good fodder grass.

Distribution. — Southern India, Rajputana, Scind and Khandeish.

[Pg 260]

Chloris virgata, *Sw.*

This grass seems to be a perennial. The stems are somewhat flattened, erect, tufted, leafy at the base and occasionally with creeping stems rooting at the lower nodes varying in length from 10 to 21 inches.

The *leaf-sheaths* are glabrous, compressed, upper sheaths somewhat inflated; mouth of the sheath is bearded with long hairs in the leaves of young branches and quite glabrous when old and in flower-bearing branches, margins are thin and membranous. The ligule is a thin narrow membranous ridge.

The *leaf-blades* are rather narrow, linear, flat, acute, glabrous when old, and with scattered long hairs in the leaves of young branches, varying in length from 2 to 9 and sometimes even 15 inches and in breadth about 1/8 inch or less.

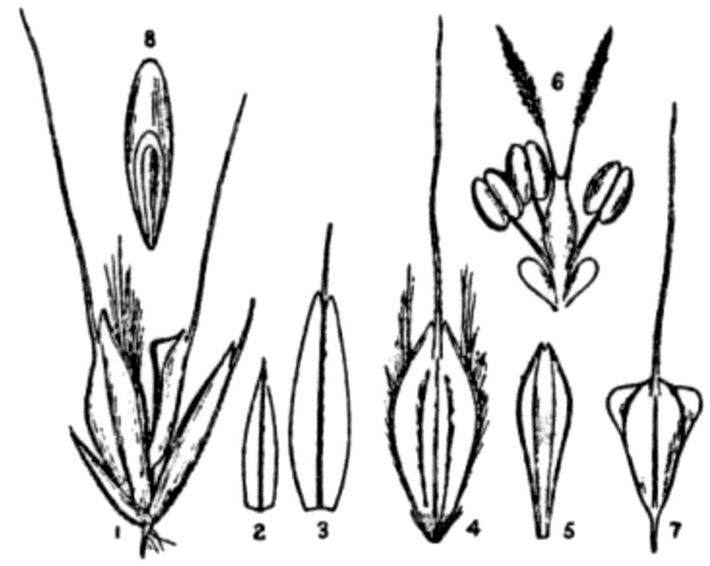

Fig. 198. — Chloris virgata.
1. Spikelet; 2 and 3, the first and second glumes; 4 and 5. the third glume and its palea; 6. lodicules, stamens and the ovary; 7. the fourth glume; 8. grain.

The *inflorescence* consists of from four to nine spikes digitately arranged on a long peduncle and the leaf-sheath enclosing the inflorescence is somewhat large and inflated. Spikes are 1 to 1-1/2 inches long with fine, angular rachis, scaberulous in the edges.

Spikelets are about 1/10 inch, 2-awned, shortly stalked and consist of only four *glumes*. The *first glume* is small lanceolate, glabrous, with the keel scaberulous, 1-nerved. The *second glume* is about one and a half times the first, oblong-lanceolate, 2-fid at the apex, glabrous, but the keel scaberulous and nerve produced between the lobes into a short scaberulous awn. The *third glume* is oblong-ovate, lanceolate, 2-fid at the apex, and awned in the sinus, awn being about 1/4 inch long bearded at the base, the margins are slightly ciliate up to about the middle and then closely ciliate with long hairs almost to the tip, but not to the tip; on the two sides of the dorsal nerve there are two shallow grooves one on [Pg 261] each

344

side, with short scattered appressed hairs; the palea is narrow ob-lanceolate, minutely 2-fid at the tip, with margins folded inward and embracing the *stamens, ovary* and the *lodicules*. Grain is narrow, trigonous, oblong, translucent and shining. The *fourth glume* is borne by a short rachilla which is about 1/3 the length of the third glume or less, shorter than the third, cuneiform, empty and awned.

This grass grows well and produces a fair amount of foliage.

Distribution. — This is not very common. So far collected only from Hosur in Salem district and Bellary district although its distribution is said to be Central and Southern India. It was found growing abundantly on old walls of houses in Poona city in 1920 and 1921.

[Pg 262]

Fig. 199. — Chloris barbata (perennial plant).

[Pg 263]

Fig. 200.—Chloris barbata.

[Pg 264]

Chloris barbata, *Sw.*

This is a very common perennial grass.

Stems are stout, tufted, geniculately ascending and erect when in flower, and some creeping and rooting at the nodes; leafy at the base and branching upwards, 1 to 3 feet; the lower internodes are 2 to 3 inches long and the upper still longer, glabrous.

The *leaf-sheaths* are glabrous, compressed laterally, open at the base and closed above, with a few scattered long hairs at the mouth, the margins thinly membranous. The *ligule* is a very narrow membrane. The *nodes* are glabrous mostly bearing tufts of leaves with compressed equitant sheaths.

The *leaf-blade* is narrow linear, flat or folded, acuminate, with long hairs on the margin towards the base, varying in length from 2 to 18 inches.

Fig. 201. — Chloris barbata.
1 to 5. the first, second, third, fourth and the fifth glume of a spikelet; 3a and 3b. the third glume and its palea; 3c. ovary, stamens and lodicules; 4a and 5a. the fourth and fifth glumes; 6. grain.

The *inflorescence* consists of five to fourteen or fifteen sessile, digitately arranged spikes, varying in length from 1-1/2 to 3 inches, on a slender peduncle; the rachis is slender minutely hairy swollen at the base.

The *spikelets* are green or purplish, 3-awned, unilaterally biseriate on the outside of the rachis, 1/10 inch excluding the awn; the *rachilla* is bearded at the base, but is shorter than the third glume and bears two barren glumes. There are five *glumes*. The *first* and the *second glumes* are lanceolate, acute, membranous, pale and 1-nerved, but the first glume is shorter than the second. The *third glume* is broadly elliptic or ovate, concave, awned, 3-nerved, with margins densely bearded above the middle and sparsely bearded dorsally on both the sides of the mid-nerve; the *palea* is oblanceolate, as long as the glume, folded inside along the margins and outside along the middle, enclosing three *stamens* and *ovary*. The [Pg 265] *fourth glume* is cuneiform, 3-nerved, awned, shortly ciliate above the middle, empty. The *fifth glume* is awned, 3-nerved, glabrous, and globose.

This grass is very widely distributed and it grows in all kinds of soils. Cattle eat it when young, but avoid it when the inflorescence is mature.

Distribution.—Throughout the plains in India, Burma and Ceylon.

[Pg 266]

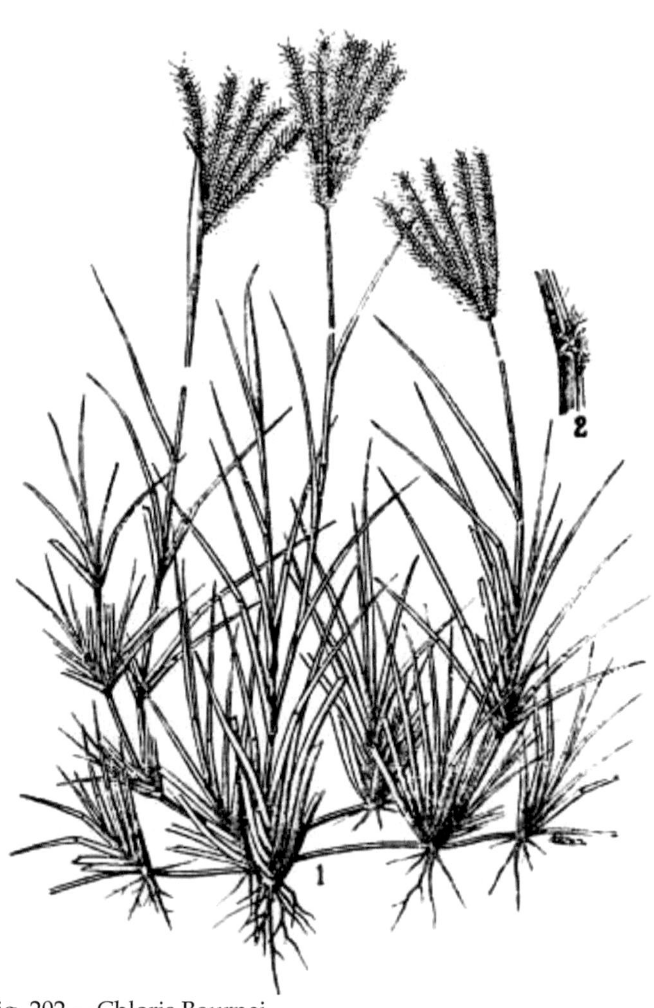

Fig. 202. — Chloris Bournei.
1. Full plant; 2. leaf showing ligule.

[Pg 267]

Chloris Bournei, *Rang. & Tad.*

This grass appears to be perennial. The stems are somewhat stout, tufted, erect or ascending geniculately from a creeping and rooting base, varying in length from 1 to 3 feet and with internodes to 6 inches becoming longer upwards.

The *leaf-sheaths* are equal to or longer than the internodes at the base, but shorter above, glabrous, compressed, distichous, bearded towards the mouth and with membranous margins. The *ligule* is a narrow membranous ridge. *Nodes* are thickened, deeply purple ringed, glabrous and the lower nodes always with a fan-like tuft of flattened leaf-sheaths and leaves.

The *leaf-blades* are linear, finely acuminate, slightly broadened and rounded at the base, keeled, the upper surface scaberulous and with a few scattered long hairs especially towards the base, smooth or slightly scaberulous below, 1 to 9 inches by 1/12 to 1/4 inch.

The *inflorescence* consists of digitately arranged spikes 1-1/2 to 4 inches long on a peduncle which is sometimes 15 inches long. *Spikes* are stout, purple-tinged, three to seven and even nine in some specimens, shortly stalked, the base of the stalk being slightly swollen and villous at the base, the rachis is slender, somewhat villous towards the base.

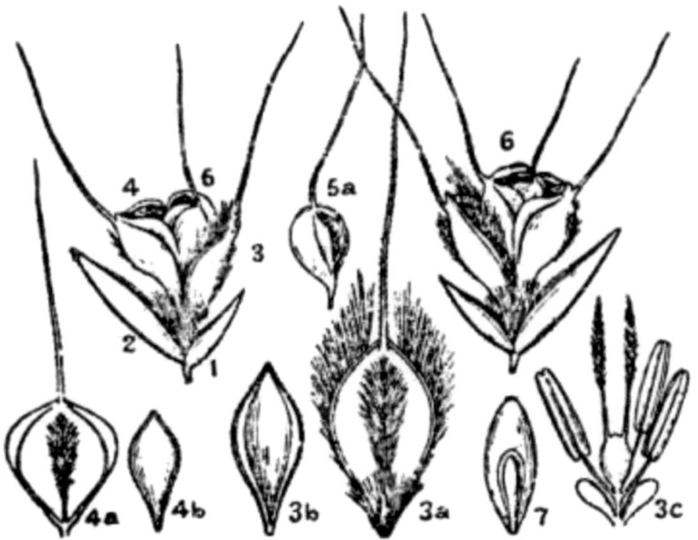

Fig. 203. — Chloris Bournei.
1 to 5. The glumes in order; 3a and 3b. the third glume and its palea;
3c. flower; 4a and 4b. the fourth glume and its palea; 5a. fifth glume;
6. a spikelet with four awned glumes; 7. grain.

The *spikelets* are about 1/8 inch excluding the awn, very shortly
pedicelled, biseriate, unilateral, disarticulating above the first two
glumes which are persistent, purplish or pale, 1- to 3-flowered, usually 3- to 4-awned and sometimes 5-awned; *awns* are purplish 3/16
to 5/16 inch long, finely scabrid. There are five or seven *glumes* in a
spikelet. The *first glume* is hyaline, purplish or pale, about 1/10 inch
long, lanceolate, sub-acuminate, 1-nerved with a scaberulous [Pg
268] keel. The *second glume* is hyaline, about one and half times as
long as the first, oblong elliptic, minutely 2-lobed at the apex, with a
minute mucro between, 1-nerved with a scabrid keel. The *third
glume* is as long as the second, awned, pale or purple, ovate or obovate, narrowed at the base and clasping the rachilla at its base, apex
shortly 2-fid with a purple dorsal awn, 3-nerved paleate; the two
marginal nerves are densely bearded with long white or purple
tinged hairs from near the base to almost the apex and the mid-
nerve also similarly bearded with long hairs on both sides, and the
base with a tuft of long hairs; the palea is as long as the glume, cori-

352

aceous obovately-cuneate, obtuse, minutely bifid, purple-tipped, with folded hyaline margins, 2-keeled; keels shortly ciliate. *Stamens* three with yellow or purple anthers, *ovary* with two feathery *stigmas* and two *lodicules*. Grain is oblong shining light reddish brown, narrowed at both ends and somewhat trigonous. The remaining glumes *fourth* to *seventh* are borne by the rachilla, thinly chartaceous, broadly obcordate or obovate, gradually diminishing in size, purple-tinged, 3- to 5-nerved, scaberulous. The fourth and fifth glumes are empty and epaleate when the spikelets are five glumed. If there are six glumes, the *fourth* bears stamens and the ovary, the *fifth* and *sixth glumes* are empty, and in spikelets of seven glumes, the third, fourth, and the fifth glumes are flower-bearing and contain grains, and the remaining two glumes are empty.

This species is a tall robust one resembling *Chloris barbata* in its inflorescence, but with larger spikelets—as large as those of *Chloris tenella*. No doubt it is closely allied to *Chloris barbata*, but differs from it by having larger spikelets that are 3- to 5-awned and 1- to 3-flowered, and the nerves being bearded throughout their length with long hairs.

Specimens of this grass were sent to Kew and Calcutta herbariums for identification and they were named *C. montana*, with which I could not agree.

So again I sent these specimens along with specimens of what I considered *C. montana* to Dr. Stapf at Kew through Mr. Gamble and Dr. Stapf wrote about these thus:— "We have not been able to match it with any of the described species of *Chloris* and Mr. Ranga Acharya will be fully justified in describing it as a new species. We have had it apart from Wight's specimen from the following collections:—(1) Sattur, November 19, 1795, sub-Andropogon barbata, Var.? Herb Rottler. (2) Ahmednagar-Miss Shattock (U.S. Dept. Agri.—received 1914). (3)Tornagallu, Bellary district, 11th August 1901 (Ex herb Ranga Acharya in Herb, Bourne No. 3594)."

Distribution.—This grass was found growing in abundance in the fields Nos. 13, 37 and 62 of the Agricultural College and in the grounds around the Forest College, Coimbatore, and was also collected in Hagari and Samalkota.

This grass grows well and is likely to prove useful, as cattle seem to like it.

[Pg 269]

Fig. 204.—Chloris montana.

[Pg 270]

Chloris montana, *Roxb.*

This is a perennial grass usually met with on dry soils. The stems are erect, tufted, geniculately ascending from a creeping base rooting at the nodes, quite glabrous, varying in length from 4 inches to 4 feet.

The *leaf-sheaths* are shorter than the internodes, flat, compressed, glabrous, with a few hairs or not at the mouth and with membranous margins; the uppermost sheath is spathiform enclosing the inflorescence when young. The *ligule* consists of only a thin ridge of short hairs densely arranged. *Nodes* are glabrous and dark-ringed, and with fan-like spreading equitant leaf-sheaths and leaves more especially when rooting.

The *leaf-blades* are narrow linear, finely acuminate, rounded at the base, glabrous throughout, folded flat inwards, 1/2 to 8 inches long, 1/16 to 1/8 inch broad.

The *inflorescence* consists of three to six (very rarely up to nine) spikes, 1 to 3 inches long, connate at the base, erect and never spreading, the peduncle is slender, long, glabrous and copiously pubescent just below the base of the connate spikes; *rachis* is angular, slender and scabrid.

Fig. 205.—Chloris montana.
1. A portion of the spike; 2. a spikelet; 3 and 4. first and second glumes; 5 and 5a. third glume and its palea; 6, 7, 8 and 9. fourth, fifth, sixth and seventh glumes; 10. lodicules, ovary and stamens; 11. grain.

The *spikelets* are about 1/8 inch excluding the awns, shortly pedicelled, unilateral, biseriate, thin and slender, 1-flowered, pale or purple tinged, disarticulating above the two lower empty glumes, which persist on the rachis, generally 4-awned, very rarely 3 or 5; *awns* are pale or purple, 1/8 to 5/16 inch; pedicel is short, angular, scaberulous with a few pilose hairs; *rachilla* is produced but is shorter than the flowering glume. There are usually six *glumes* in a spikelet and very rarely five or seven glumes; of these the first two *glumes* are hyaline, empty, awnless; the third is flower-bearing and the rest empty, thinly coriaceous and awned. The *first glume* is white or lightly purplish, small, about 1/16 inch long, lanceolate, [Pg 271] finely acuminate, 1-nerved, and with scabrid keel. The *second glume* is twice the first glume in length, oblong-lanceolate, finely acuminate, 1-nerved. The *third glume* is broadly oblong, chartaceous, 3-nerved, bearded with long hairs along the margins from a little above the base, and with a tuft of hairs at the base and an awn at the apex; the palea is oblong, a little smaller than the glume, folded along the margins. There are three *stamens* with pale yellow anthers. The *styles* are white with purple *stigmas*. *Lodicules* are narrowly cuneate. The *fourth* and the *fifth glumes* are small, epaleate, empty, oblong, cuneate, 3-nerved, awned. The *sixth glume* is very small, cuneate, awned.

Distribution.—In the districts forming the Coromandel Coast and also Gangetic plains and Ceylon.

[Pg 272]

37. Eleusine, *Gaertn.*

These are annual or perennial grasses. Leaves are long or short. The spikelets are sessile, 3 to 12 flowered, 2 to 3-seriate, secund, laterally compressed and forming digitate whorled or capitate spikes, not joined at the base; rachilla continuous between the flowering glumes. The glumes in a spikelet are few to many, keeled. The first two glumes are subequal or unequal, persistent; the first glume is 1-nerved and the second glume is 1- to 7-nerved. The flowering glumes are 3-nerved, paleate; palea is complicate; keels are strong, scabrid or ciliate. Lodicules are two, cuneate. Anthers are short. Styles distinct and short. Grain is free, rugose, and the pericarp is hyaline and loose.

KEY TO THE SPECIES.

- Spikelets pointing upward at an acute angle with the rachis of the spike.
    - Spikes 1 to 5 inches long, digitate, erect. 1. E. indica.
    - Spikes 1/6 to 1/4 inch or a little more, capitate, spreading. 2. E. brevifolia.
- Spikelets spreading at right angles with the rachis of the spike, spreading or erect. 3. E. ægyptiaca.

[Pg 273]

Eleusine indica, *Gaertn.*

This is a tufted annual grass with short, erect, somewhat compressed, glabrous stems, 1 to 2 feet high.

The *leaf-sheaths* are compressed, distichous, ciliate. The *ligule* is a ridge of hairs.

The *leaf-blades* are narrow-linear, as long as the stem, glabrous or with a few scattered hairs near the mouth, acuminate, base not contracted, 12 to 20 inches long and 1/8 to 1/6 inch broad.

The *spikes* are elongate, digitate, 2 to 7, 2 to 5 inches long, all in a terminal whorl and sometimes with one or two lower down, and with the axils glandular and hairy; the *rachis* is slender and dorsally flattened.

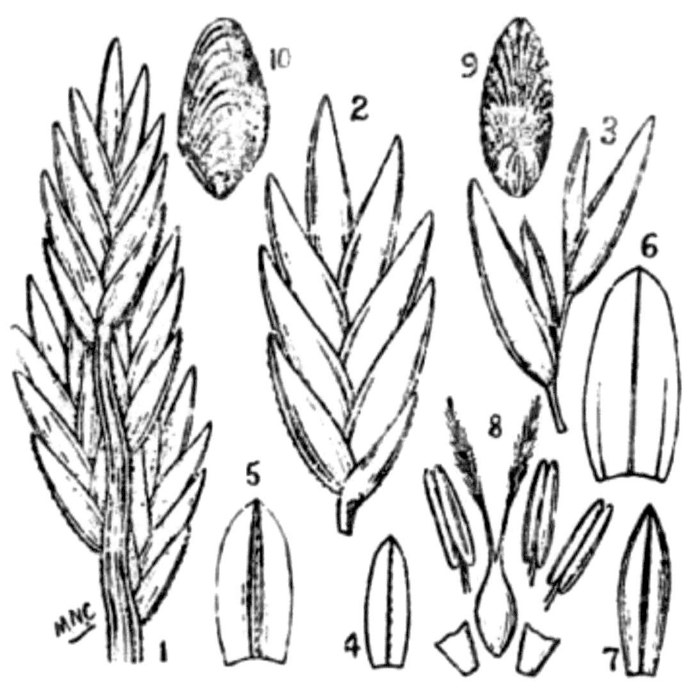

Fig. 206.—Eleusine indica.
1. A portion of the spike; 2. a spikelet; 3. flowering glumes and their palea with the rachis; 4 and 5. the first two glumes; 6 and 7. flowering glume and its palea; 8. the ovary, stamens and the lodicules; 9 and 10. grain.

The *spikelets* are variable in size, 1/12 to 1/6 inch, 3 to 5, rarely 6-flowered, quite glabrous, biseriate, pointing upward at an acute angle with the rachis. All the glumes are more or less membranous. The *first glume* is small, oblong-ovate or oblong, 1-nerved with a scabrid keel. The *second glume* is twice the size of the first, ovate-oblong, 3-nerved, rarely 3- to 7-nerved, glabrous, shortly mucronate at the acute apex. The *third glume* and the succeeding flowering glumes are larger than the second, ovate-oblong, subacute, 3-nerved and paleate; *palea* is shorter than the glume, glabrous. *Stamens* are three. *Lodicules* are small and cuneate. The grain is oblong, obtusely trigonous, broadly and shallowly grooved dorsally with concentric minute tubercled ridges covered with a loose pericarp.

This grass is fairly common in somewhat wet places in the plains and low hills.

Distribution. — Throughout India and Ceylon.

[Pg 274]

Eleusine brevifolia, *Br.*

This is an annual grass. Stems are creeping and spreading from the root, and ascending from a decumbent base, generally slender and small, but sometimes large and proliferously branched, leafy, 3 to 7 inches long.

The *leaf-sheath* is compressed and glabrous. The *ligule* is a very short membrane, ciliate at the margin or obsolete.

The *leaf-blade* is linear, acute, with a subcordate or rounded base 1/2 to 2 inches long and 1/8 to 1/6 inch broad.

The *spikes* are usually many, sessile and crowded in globose heads, varying in diameter from 1/3 to 2/3 inch.

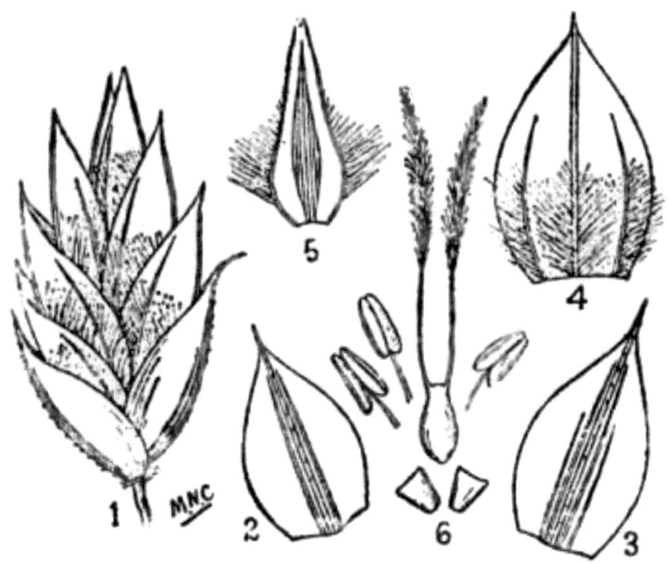

Fig. 207.—Eleusine brevifolia.
1. A spikelet; 2 and 3. the first and the second glumes; 4 and 5. the third glume and its palea 6. lodicules, ovary and stamens.

Spikelets are sessile, biseriate, ovate-oblong, 1/8 to 1/6 inch long, 4- to 10-flowered. The *first two glumes* are membranous, ovate-oblong, glabrous, acuminate and shortly awned, the *first glume* is shorter than the second, 1- to 3-nerved, the *second glume* is longer than the first, 3- to 5-nerved, and the nerves are very close to the middle one in the keel. The *third* and the succeeding *glumes* are ovate, cuspidately acuminate, 3-nerved, nerves villous below the middle and paleate; *palea* is oblong, lanceolate, truncate and minutely 2-toothed, keels villous below the middle. *Anthers* are small. *Lodicules* are also small and cuneate. *Styles* are long and slender. Grain is orbicular to ovate, concavo-convex, red-brown, and transversely rugose.

This grass is usually found in somewhat damp situations all over the Presidency, though somewhat local in its distribution.

Distribution.—Sandy shores of the Coromandel and Carnatic coasts.

[Pg 275]

Fig. 208.—Eleusine ægyptiaca.

[Pg 276]

Eleusine ægyptiaca, *Desf.*

This grass is an annual with erect or creeping branches. Stems are
erect or prostrate, compressed, smooth, spreading and rooting at the

nodes, 6 to 18 inches long. Nodes are thickened and sometimes proliferous.

The *leaf-sheath* is compressed and glabrous. The *ligule* is short and membranous.

The *leaf-blade* is linear, tapering to a fine point, flat, glaucous, glabrous or hairy, 1 to 6 inches long and 1/12 to 1/6 inch, wide.

Fig. 209. — Eleusine ægyptiaca.
1. Front and back views of a portion of spike; 2. a spikelet; 3 and 4. the first and the second glumes; 5 and 6. flowering glume and its palea; 7. ovary and anthers.

Spikes are digitate, 2 to 6, 1/2 to 1-1/2 inches long. *Spikelets* are flat, densely crowded on one side of the floral axis, spreading at right angles, 3- to 5-flowered, *glumes* five to seven. The *first glume* is ovate acute. The *second glume* is equal to the first or slightly longer, broadly ovate, awned. The flowering *glumes* are ovate, mucronate or awned, paleate; *palea* is shorter than the glume, ovate-oblong, obtuse or 2-fid. *Anthers* are small. Grain is reddish, rugose and sub-globose.

This is a very common grass occurring as a weed in cultivated fields and in open places. It is a well-known fodder grass.

Distribution. — Throughout the plains in India and Ceylon.

[Pg 277]

38. Dinebra, *Jacq.*

These are leafy annual grasses. The inflorescence is a narrow pyramidal raceme of slender, spreading or deflexed spikes. Spikelets
are small, biseriate and crowded on one side of the spike and not
jointed at the base; rachilla is slender, jointed and produced beyond
the flowering glumes and bearing an imperfect glume. There are
four to five glumes. The first two glumes are the longest, lanceolate,
1-nerved, keeled and awned. The second glume is slightly longer
than the first. The third and the fourth glumes are very small, hyaline, broadly ovate, 1-nerved. Lodicules are present. Stamens are
three and anthers didymous and small. Grain is narrowly ovoid and
trigonous.

[Pg 278]

Fig. 210.—Dinebra arabica.
1. Full plant; 2. leaf showing the ligule.

[Pg 279]

Dinebra arabica, *Jacq.*

This grass is an annual with stems erect or with a geniculate base, tufted, slender or stout; some of the lower nodes of the geniculate part of the stems bear roots; the internodes are green or purple tinged and glabrous.

The *leaf-sheath* is thin, somewhat loose, usually glabrous, rarely sparsely hairy. The *ligule* is a short membrane irregularly cut at the top. The *nodes* are glabrous.

The *leaf-blade* is linear, very finely acuminate, rough on both the surfaces, thinly and very sparsely hairy; the base of the blade is contracted and purple tinged towards the margin, midrib is prominent with three or four main veins on each side; the margins are very finely, closely serrate.

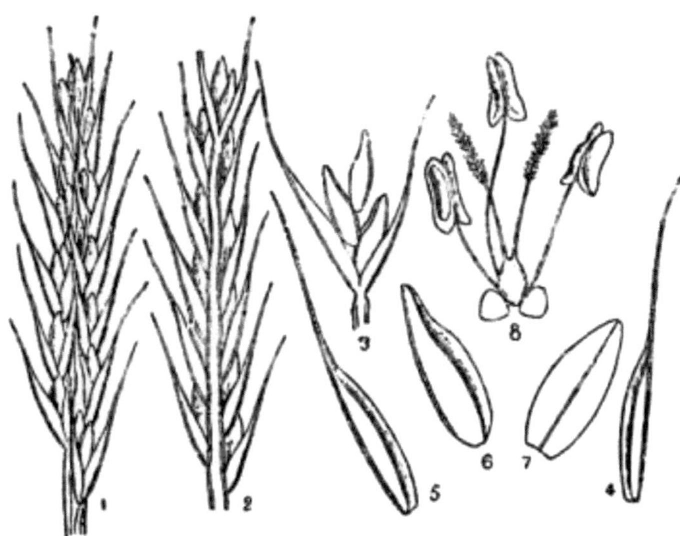

Fig. 211.—Dinebra arabica.
1 and 2. The front and back view of a portion of a spike; 3. spikelet;
4, 5 and 6. the first, second and third glumes; 7. palea of the third
glume; 8. lodicules, ovary and stamens.

The *inflorescence* is a long erect narrow pyramidal panicle varying in length from 2 to 16 inches; the lower branches sometimes bear several spikes and attain 6 inches in length; the *peduncles* are short or long, purple tinged and the main *rachis* is smooth except at the top, angular and grooved. The *spikes* are numerous, greenish or purple tinged, slender, erect or spreading or sometimes deflexed, opposite, alternate or in fascicles of two to four varying in length from 1/4 to 2-1/2 inches; the *rachis* of the spike is trigonous, flattened out ventrally and with a ridge on the ventral side and the margins are scabrid.

The *spikelets* are few to many in a spike, alternate, closely imbricating, sessile, about 1/6 inch long including the awns, usually three flowered, rarely less or four flowered; the *rachilla* is very slender, jointed at the base, produced and jointed between the flowering glumes.

There are usually five *glumes* in a spikelet and in some four or six. The *first* and the *second glumes* are lanceolate narrowed into [Pg 280] short stiff awns, equal or the second a little longer, hyaline glabrous, strongly keeled about 1/6 inch long or a little less. The *succeeding glumes* third, fourth and fifth are very much shorter than the first two glumes, about 1/10 inch or less, ovate-oblong, subacute, white, membranous with a strong greenish nerve along the keel and two short ones close to the margin, paleate; *palea* is shorter than the glume, membranous, oblong-obtuse, minutely 2-toothed, 2-nerved and 2-keeled. *Stamens* are three with small anthers. *Stigmas* are white when young and purple when mature. *Lodicules* are very minute. The grain is pale, brownish yellow, ellipsoidal-oblong, subacute, trigonous, rough and never smooth, with a shallow groove on the dorsal side; the embryo is about one-third the length of the grain.

This grass grows abundantly in cultivated dry fields all over the Presidency. The spikes when mature become very rough and give an acid taste. Cattle greedily eat this grass when young, but when old and in full flower some cattle do not like it so much.

Distribution.—Throughout the Presidency in the plains. Also occurs in Afghanistan and westward to Senegal.

[Pg 281]

39. Leptochloa, *Beauv.*

These are tall slender annual grasses. Spikelets are very small, compressed, 1- to 6-flowered, sessile or shortly pedicelled, alternate and unilateral on the branches of a panicle; the rachilla is produced between the flowering glumes, jointed at the base. There are 3 to 8 glumes. The first two glumes are unequal, oblong or lanceolate, 1-nerved. The third and the succeeding ones are broadly ovate, 3-nerved, paleate. Lodicules are two. Stamens are three. Grain is sub-globose, oblong or trigonous, closely invested by the glume and its palea.

[Pg 282]

Leptochloa chinensis, *Nees.*

This is a tall annual grass. Stems are erect or geniculately ascending from a creeping root-stock, varying in length from 2 to 4 feet.

The *leaf-sheath* is smooth, loose, the lower often broad and open. The *ligule* is a short hyaline lacerated membrane.

The *leaf-blade* is narrowly linear, finely acuminate, somewhat coriaceous, glabrous, 6 to 18 inches long and 1/6 to 1/4 inch broad.

The *inflorescence* is a contracted panicle, 6 to 18 inches long with spreading or suberect, alternate or opposite spikes which are capillary and vary from 2 to 4 inches in length.

Fig. 212.—Leptochloa chinensis.
1. A portion of the spike; 2 and 3. the first and the second glume; 4 and 5 the flowering glume and its palea; 6. the stamens and the ovary.

The *spikelets* are small, shortly stalked, 4- to 8-flowered, 1/10 to 1/6 inch with the *rachilla* produced between the flowering glumes. The *first glume* is small, oblong, obtuse or apiculate. The *second glume* is similar to the first but twice as long as the first glume. The *third glume* and the succeeding flowering glumes are ovate-oblong, obtuse or apiculate, with sub-marginal lateral veins; *palea* are broadly oblong with silkily ciliate keels. *Anthers* are usually very small.

Grain is oblong, obtusely trigonous, or concavo-convex, red-brown and rugulose on the ventral side.

This grass is very common amidst paddy in wet lands and in wet situations.

Distribution.—Throughout India and Ceylon in wet places. Also in China, Japan and Australia.

[Pg 283]

CHAPTER X.
TRIBES VII AND VIII – FESTUCACEÆ AND HORDEÆ.

Festucaceæ is of minor importance as it is not well represented in the South India. Only about half a dozen genera occur and most of them on the hills. The spikelets are usually 2- or more-flowered, pedicelled and in panicles, open or contracted. The rachilla is produced beyond the flowering glumes and articulate at the base just above the empty glumes.

- Inflorescence a raceme, spikelets 2- to 3-flowered, turbinate; glumes single-awned. 40. Pommereulla.
- Inflorescence paniculate, spikelets few or many-flowered, glumes many-nerved and many-awned. 41. Pappophorum.
- Inflorescence various, spikelets 2- to many-flowered, flowering glumes 1- to 3-nerved entire, empty glumes shorter than the lowest flowering glume, grain very minute. 42. Eragrostis.

Hordeæ is also a minor tribe and is represented by only one genus in South India.

The spikelets are one-or more-flowered, sessile, 1- or 2-seriate on the rachis, and somewhat sunk in cavities; the rachilla is jointed at the base and is produced beyond the flowering glumes, glumes awned or not.

- Spikelets 1- to 3-flowered, first glume very minute or wanting, second as long as the hyaline, third spike compressed, solitary. 43. Oropetium.

[Pg 284]

40. Pommereulla, *Linn. f.*

This is a short, stout, creeping perennial grass. Spikelets are 2- to 3-flowered, distichously racemed, narrowly turbinate, villous. Glumes are 5 to 7 in a spikelet. The first two glumes are narrow, membranous, persistent, the first glume being 1-nerved and shorter than the second which is 3- to 5-nerved. The third and the fourth glumes embracing the fifth and the sixth are empty, flabelliform, 4-lobed, and dorsally shortly awned. The fifth, sixth and the seventh are cuneate, obovate and 3-lobed, palea ovate, acute, and pubescent. Lodicules are two and membranous. Stamens are two to three with small anthers. Grain is oblong, compressed and free.

[Pg 285]

Fig. 213.—Pommereulla Cornucopiæ.

[Pg 286]

Pommereulla Cornucopiæ, *Linn. f.*

This is a short, stout perennial grass with stems rooting at the nodes; branches are flat, short, densely leafy, 2 to 6 inches long.

The *leaf-sheaths* are smooth, equitant with thinly membranous margins. The *ligule* is a ciliated ridge.

The *leaf-blade* is flat, linear, distichous, coriaceous, rounded at the tip, margins sparsely ciliate, 1 to 2-1/2 inches long.

The *inflorescence* is a terminal raceme, 1/2 to 2 inches long, half hidden by the uppermost leaf-sheath, the peduncle is flattened and 1 to 2-1/2 inches long; rachis is also flattened with a tuft of long silky hairs at the base.

Fig. 214.-Pommereulla Cornucopiæ.
1. A leaf; 2. inflorescence; 3. spikelet; 4 and 5. the second and the first glume; 6 and 7. the third and the fourth glume; 8 and 9. the fifth flowering glume and its palea; 10 and 11. grain.

The *spikelets* are shortly pedicelled or sessile, dorsally compressed, cuneiform, about 1/3 inch, glistening, villous, not articulate at the base, 2- to 3-flowered, rachilla is narrowed downwards, resembling a callus and villous, jointed at the acute base above the empty glumes, and crowned with broad obconic empty awned glumes. The spikelets have usually seven, rarely eight glumes. The *first* and the *second glumes* are narrow, membranous, glistening, empty and persistent and the others are coriaceous with membranous margins. The *first glume* is linear or linear-lanceolate, acuminate, 1-nerved, scaberulous along the nerve. The *second glume* is longer than the first, oblong-lanceolate, acuminate, narrowed towards the base, inserted much above the first glume and embracing the rachilla, 3-nerved, scaberulous along the mid-nerve at the base only. The *third* and *fourth glumes* are half-amplexicaul, empty, epaleate, flabelliform, 4-lobed, 7-nerved, shortly awned at [Pg 287] the back, villous; the side lobes are acuminate or aristate and the central lobes are shortly awned. The *fifth*, *sixth* and *seventh glumes* are obovate-cuneate, 7- to 9-nerved, paleate, flower-bearing and 3-lobed, the side lobes are acuminate and the central lobe is bifid and dorsally awned; palea is ovate-acute, 2-nerved and ciliolate. The *eighth glume*, if present, is neuter and imperfect, 3-lobed and shortly awned. *Lodicules* are minute. *Stamens* are two or three with small anthers. Grain is oblong, compressed, reddish brown.

This grass generally grows in gravelly and somewhat alkaline soils. So far this has been noticed and collected in Chingleput and Nellore districts.

Distribution.—Mysore and the Carnatic, and Ceylon.

[Pg 288]

41. Pappophorum, *Nees.*

This is a perennial grass. Spikelets are contracted spiciform panicles, 1- to 3-flowered, rachilla is jointed at the base. There are 5 to 7 glumes in the spikelet. The *first* and the *second glumes* are membranous, keeled 3- to many-nerved, persistent. The *third* and the fourth glumes are much shorter (excluding the awns) than the first two, coriaceous, orbicular, concave, obscurely many-nerved, cleft into nine or more equal or alternately longer long-ciliate erect awns. The fourth and the subsequent glumes are imperfect and they get gradually smaller and smaller, the last glume being represented only by a rudimentary glume with three awns. Lodicules are dolabriform and two. Stamens are three. Styles are free. Grain is obovoid or oblong, free.

[Pg 289]

Fig. 215.—Pappophorum elegans.

[Pg 290]

Pappophorum elegans, *Nees.*

This grass is perennial with wiry roots. Stems are erect ascending from a swollen woody base, thinly hairy and rarely glabrous, pale green and sometimes with red blotches, wiry, varying in length from 1 to 3 feet.

The *leaf-sheath* is thinly pubescent, some hairs being minutely gland-tipped.

The *leaf-blade* is narrow, linear-lanceolate, sharply acuminate, covered both above and below with hairs, many of which being minutely gland-tipped, convolute when young. The *ligule* is a ridge of hairs. Nodes are pubescent.

The *inflorescence* is a panicle with short branches, 1 to 3 inches long, rachis is pubescent; peduncle is 2 to 4 inches long, pubescent. The *spikelets* are pale green, sometimes purple tinged and appearing white when mature, softly pubescent, about 1/4 inch long including the awn; the rachilla is produced and disarticulates above the two lower glumes.

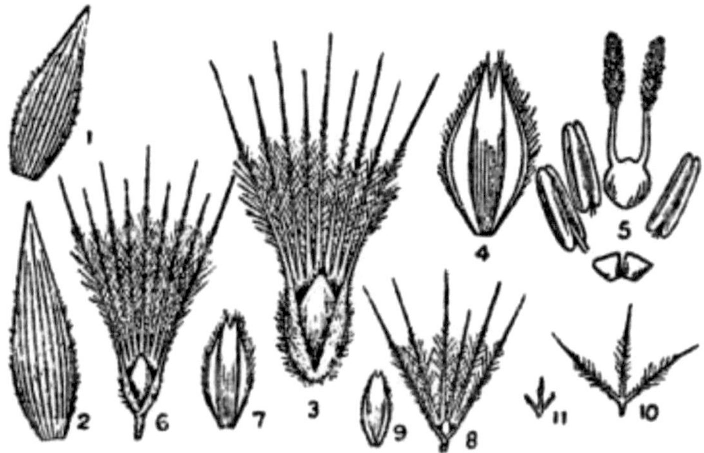

Fig. 216.—Pappophorum elegans.
1 and 2. The first and second glumes; 3. the third glume and its palea; 4. palea of the third glume; 5. lodicules, stamens and ovary; 6

and 7. fourth glume and its palea; 8 and 9. fifth glume and its palea; 10 and 11. sixth and seventh glumes.

There are 6 or 7 glumes in the spikelet. The *first glume* is lanceolate, acute, softly hairy, usually 9-nerved, or varying from 7 to 12 (some nerves do not reach the apex), about 1/4 inch long. The *second glume* is similar to the first but a little longer and both the glumes have broad hyaline margins. The *third glume* is broadly orbicular, concave, sub-chartaceous, 9-nerved, densely villous and with a tuft of hairs at the base where it joins the rachilla, cleft into 9 awn-like lobes, bisexual and paleate; the awns are alternately long and short, subulate, plumose in the lower half and scabrid above, the palea is oblong-ovate, sub-chartaceous, with two pubescent keels, bifid at the apex, and with 3 purple anthers. The *ovary* is ovoid or ovoid-oblong, with two white stigmas. *Lodicules* are two, small cuneate or quadrate. Grain ovoid or ovoid-oblong. The *fourth* [Pg 291] *glume* is similar to the third glume but smaller, paleate with rudimentary anthers and two fleshy lodicules. The *fifth, sixth* and *seventh glumes* are imperfect and gradually decreasing in size, and with awns varying in number from 5 to 8, 3 to 5, and 1 to 3, respectively, minutely paleate or not.

This grass grows well in black cotton and rich loamy soils and is a hardy one. Cattle seem to eat this grass.

Distribution.—Fairly common in the plains in the Deccan districts and in the Coromandel coast districts.

[Pg 292]

42. Eragrostis, *Beauv.*

These are slender, glabrous, annual or perennial grasses. Stems are usually erect or geniculately ascending, very rarely prostrate. Leaves are narrow. Inflorescences are open or contracted panicles, rarely spikes. Spikelets are usually strongly laterally compressed, 2, to many-flowered and not articulate at the base; rachilla is tough and persistent, jointed above the empty glumes and in some also between the flowering glumes, not produced beyond the last glume. Glumes are many, broad, obtuse, acute or mucronate, never awned, dorsally rounded and keeled; the first and the second glumes are much shorter than the spikelet, equal or unequal, empty, persistent or separately deciduous, 1-nerved or the second 3-nerved, usually membranous. Flowering glumes are imbricating, at length deciduous from the rachilla, 3-nerved, all bisexual or the uppermost and rarely the lowest imperfect, ovate to lanceolate, membranous to chartaceous, usually glabrous, the lateral nerves short not reaching the mid nerve; palea are broad, membranous, deciduous with its glume or persistent on the rachilla with two ciliate smooth or scabrid keels. Stamens are three rarely two. Ovary is glabrous with two styles ending in plumose stigmas. Grain is minute, globose, obgloboid or obovoid, free in the glume and the palea.

KEY TO THE SPECIES.

- A. Spikelets panicled.
  - B. Rachilla of spikelets more or less jointed and breaking up from above downwards.
    - Panicle more or less contracted and margin of flowering glumes not ciliate.
      - Spikelets 1/20 to 1/6 inch long; grain obovoid; stamens 2; panicle narrow interrupted, 6 to 18 inches long 1. E. interrupta.
  - BB. Rachilla of spikelets tough, persistent; flowering glumes falling away from base upwards.
    - C. Spikelets pedicellate.
      - Spikelets flat, ovate-elliptic or oblong, lateral nerves of flowering glumes very prominent and straight, almost percurrent; palea deciduous with their glumes 2. E. amabilis. [Pg 293]
      - Spikelets less compressed, linear or linear-oblong; lateral nerves less prominent; not fascicled, long pedicellate and divaricate when ripe.
        - Leaf margins without glands. Spikelets versa-

- tile, narrow, linear 1 inch or more long, branches of panicle solitary 3. E. tremula.
- Leaf margins glandular.
  - First glume 1-nerved and second glume 3-nerved 4. E. major.
  - First glume and second glume 1-nerved 5. E. Willdenoviana.

  - Spikelets small, 1/4 inch or less, branches of panicle whorled 6. E. pilosa.
- CC. Spikelets sessile and jointed on the very short densely crowded branchlets of a tall, narrow raceme like panicle, deciduous, acute, much compressed, imbricate and secund 7. E. cynosuroides.
- AA. Spikelets in a long terminal spike. Spikelets distichously spreading, secund, keels of palea winged 8. E. bifaria.

[Pg 294]

Eragrostis interrupta, *Beauv.*

(*Var. Kœnigii,* Stapf.)

This is a tall grass, annual or perennial, with erect stems 1 to 3 feet or more.

The *leaf-sheath* is glabrous and close. The *ligule* is a short, fimbriate membrane. *Nodes* are glabrous.

The *leaf-blade* is narrow, flat, acuminate, glabrous on both sides, 3 to 10 inches long.

The *panicle* is erect, narrow, contracted, with branches in pseudo-whorls and varying in length from 6 to 18 inches, branches are slender, filiform, two or more arising from the same level, 1 to 3 inches long.

The *spikelets* are small, pedicellate, smooth, usually 6 to 14-flowered, pale but often tinged with red, the rachilla is jointed between the flowering glumes, and breaks away from above downwards. The empty *glumes* are very small, subequal, ovate-oblong, hyaline, obtuse and 1-nerved. Floral *glumes* also are small but slightly longer than the empty ones, ovate-oblong, obtuse and paleate, palea is linear-oblong with smooth or scabrid keels. *Stamens* are two with small anthers. Grain is obovoid.

Fig. 217.—Eragrostis interrupta. Var. Kœnigii.
1. Two spikelets; 2 and 3. empty glumes; 4. empty glumes with two flowering glumes and their palea; 5. flowering glumes and palea; 6. ovary and two stamens; 7. grain.

This grass is a very variable plant and has a few varieties. The one described above is Var. *Kœnigii* Stapf., and this is the one that occurs very widely. The other two varieties which occur very rarely are (1) *diarrhena* Stapf. and (2) *tenuissima* Stapf. The former is a tall plant with very narrow panicle and spikelets and the latter either tall or short and with a panicle bearing very slender divaricate branches.

This grass usually occurs in clayey soils especially on the bunds and in the paddy fields.

Distribution. — Throughout India, Burma and Ceylon. Also in tropical Asia and Africa.

[Pg 295]

Eragrostis amabilis, *W. & A.*

This is an annual tufted grass with slender, glabrous, erect or geniculately ascending stems, 6 to 18 inches, leafy chiefly at the base.

The *leaf-sheath* is glabrous and smooth. The *ligule* is absent or very obscure.

The *leaf-blade* is lanceolate-linear or linear, narrowed from a broad subcordate base to an acute tip, smooth and flat.

The *panicle* is ovoid-oblong or oblong, open or contracted, sparingly branched; branches are filiform, solitary, ramifying from near the base; rachis and nodes are glabrous.

Fig. 218.—Eragrostis amabilis.
1. A portion of a branch with spikelets; 2. a single spikelet; 3 and 4. empty glumes; 5. and 6. a flowering glume and its palea; 7. lodicules, stamens and ovary; 8. grain.

The *spikelets* are ovate-oblong or linear-oblong, pale or purplish 1/6 to 1/2 inch, up to 50-flowered, rachilla is tough with very short internodes. The glumes are very closely and distichously imbricating (and hence spikelets are pretty); the *empty glumes* are subequal, ovate-lanceolate, acute or cuspidately acuminate, 1-nerved, 1/25 to 1/16 inch long. *Flowering glumes* are broadly ovate or suborbicular, mucronulate, punctulate, with the lateral nerves equidistant from the margins and the median nerve, and produced far up towards the median nerve; palea is broad, shorter than its glume, deciduous with it, and with winged and scabrid keels. *Stamens* are three. Grain is obovoid-ellipsoid, smooth, laterally compressed, reddish-brown.

This grass is abundant in wet places on the hills and fairly common in the plains though not abundant.

Distribution.—Throughout India and Ceylon.

Fig. 219. — Eragrostis tremula.

Eragrostis tremula, *Hochst.*

This is an elegant annual grass. Stems are tufted erect or some-times geniculately ascending, branching freely, 6 inches to 3 feet.

The *leaf-sheath* is smooth, glabrous, shorter than the internodes, becoming purplish when dry. The *ligule* is a ridge of short hairs.

The *leaf-blade* is linear-lanceolate, tapering to a fine point, rigid, glabrous or sparsely hairy, but with prominent white hairs near the mouth of the sheath at the base, 1 to 10 inches long and 1/12 to 3/16 inch broad, the base is rounded and the margin eglandular and very finely serrate.

The *inflorescence* is a large, effuse, nodding, pyramidal or oblong panicle, much branched, the peduncle being as long as the rest of the plant; branches are slender, solitary, suberect, drooping, rather angled, scaberulous, 3 to 7 inches long with very fine capillary branchlets; all the axils of the branches and branchlets have long white hairs.

Fig. 220.—Eragrostis tremula.
1. Spikelet; 2 and 3. the first and the second glume; 4 and 5. flowering glume and its palea; 6. stamens, ovary and lodicules.

The *spikelets* are linear, narrowed upwards, glabrous, flattened pale green or purple tinged, few to 70-flowered; pedicels are slender and capillary, longer or shorter than the spikelets; rachilla is zigzag and glabrous. The *first two glumes* are subequal, ovate, acute, one-nerved, keel obscurely scaberulous, membranous. The [Pg 298] *third* and the succeeding *flowering glumes* are ovate, obtuse, as long as the second glume or slightly longer, sub-chartaceous, glabrous, three-nerved; palea is shorter than the glume, curved obovate oblong and persistent on the rachilla. *Stamens* are three with small anthers. Style branches are two. *Lodicules* are minute. Grain is nearly globose, compressed on one side, obscurely rugulose.

This grass is not very widely distributed although it occurs in some parts of the Presidency. It is common on the West Coast in sandy places.

Distribution.—From the Punjab to Bengal and Burma and Southward to Carnatic. Also said to occur in Afghanistan and Tropical Africa.

[Pg 299]

Fig. 221.—Eragrostis major.

[Pg 300]

Eragrostis major, *Host.*

This is an annual tufted grass. Stems are erect or geniculately ascending, usually short, leafy and branched below, glabrous and shining, 1/2 to 2 feet long.

The *leaf-sheath* is glabrous, striate, shorter than the internodes, keeled with tubercles or glands on the keel and also on some of the smaller nerves on the sides, and bearded with long white hairs externally at the mouth. The nodes are glabrous purple, shining and with a glandular ring below. The *ligule* is a ridge of long hairs.

The *leaf-blade* is linear-lanceolate or linear, tapering to a fine point, glabrous, flaccid, margins finely serrulate and glandular, base rounded, varying in length from 1/2 to 10 inches and in breadth 3/16 to 7/16 inch; the midrib is prominent and with a row of glands beneath and there are 3 to 5 lateral nerves on each side of the mid-nerve.

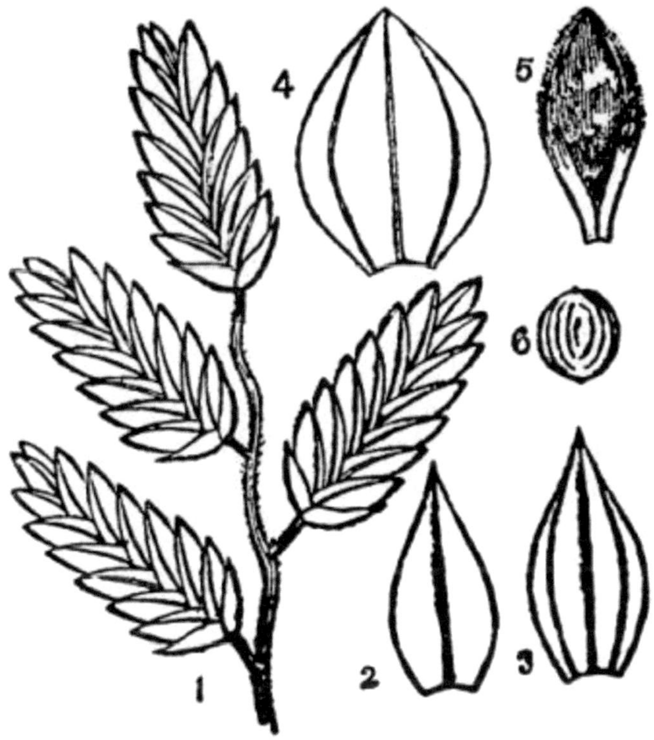

Fig. 222.—Eragrostis major.
1. A branch with spikelets; 2 and 3. empty glumes; 4 and 5. flowering glume and its palea; 6. grain.

The *panicle* is ovate or ovate-oblong, on a short, smooth peduncle, usually open and stiff; branches are usually many, sub-solitary or fascicled, spreading or suberect, capillary, stiff, again branching from near the base and about 3 inches long; *rachis* is angular, with glands and tufts of sparse white hairs at the angles of branches and branchlets.

Spikelets are linear to ovate-oblong, compressed, pale or green, sometimes purple tinged at the base, few to 40-flowered and occasionally up to 70-flowers, 1/8 to 1 inch.

The *empty glumes* are subequal or the first is a little shorter, ovate, acute, membranous, keeled, and sometimes the keels with glands; the *first glume* is usually one-nerved (rarely obscurely one- to three-nerved) and the *second glume* is three-nerved.

[Pg 301]

The *flowering glumes* are broadly ovate, oblique, obtuse, sometimes with a minute mucro, sub-chartaceous, punctulate, strongly three-nerved, paleate, about 1/12 inch long; palea is shorter than the glume, curved, obovate-oblong, keels ciliolate and persistent. *Stamens* are three with very small pale yellow anthers. Stigmas are two and white. *Lodicules* are very small. Grain is globose reddish brown, minutely and obscurely lineolate.

This grass is a very common weed occurring in cultivated dry fields all over this Presidency.

Distribution. — Throughout India and Ceylon in the plains and low hills. Occurs also in tropical and sub-tropical parts of Asia and Africa.

[Pg 302]

Fig. 223. — Eragrostis Willldenoviana.

[Pg 303]

390

Eragrostis Willdenoviana, *Nees.*

This is a tufted annual. Stems are leafy at the base, erect or genic-ulately ascending, slender but rigid, varying in length from 4 to 18 inches.

The *leaf-sheath* is smooth, cylindric, glabrous, outer margin ciliate; tufts of long hairs are present at the sides of the margin of the sheath, just outside close to the hyaline patch. The *ligule* is a fringe of short white hairs. The *nodes* are greenish or with a tinge of pur-ple, glabrous and with a glandular ring below.

The *leaf-blade* is lanceolate-linear, pointed, flat, rigid, the margin is very minutely serrulate, glandular and occasionally also with fine long hairs; the upper surface is somewhat rough, the lower smooth and both with fine long scattered hairs or glabrous.

Fig. 224. — Eragrostis Willdenoviana.
1. Spikelets; 1a. 1st glume; 2 and 2a. the second glume; 3 and 3a. the flowering glume; 4. palea of the flowering glume; 5. lodicules, sta-mens and the ovary; 6. grain.

The *inflorescence* is a stiff open panicle, ovate to oblong, 2 to 4-1/2 inches long on a slender, terete, glabrous peduncle; the main *rachis* is angular, slender with glandular scars, a little below the attach-ment of the branches; the branches are capillary, grooved stiff and spreading with small glandular scars just above the node. The *spike-lets* are elliptic-oblong to linear, 1/8 to 3/4 inch by about 1/20 inch, greenish or tinged with purple, few to about 25 (or sometimes even up to 42) glumed, pedicellate; pedicel is capillary, grooved and angular, with a glandular ring about the middle, spreading sometimes at right angles, rachilla is persistent.

Empty *glumes* are unequal. The first *glume* is hyaline very small, nerveless or one-nerved, subacute or subobtuse; the second *glume* is much longer than the first glume, ovate-oblong subacute, keeled, membranous and one-nerved. *Flowering glumes* vary from about 12 to 30 and in some well grown plants as many as 42, broadly ovate, obtuse or subacute, rigidly membranous, three-nerved (one median [Pg 304] and two marginal) glabrous, keeled and keels are scaberulous near the apex; palea is oblong linear, a little curved, persistent, a little smaller than the glume, two-keeled; there are three *stamens* with small purplish anthers and two small *lodicules*. The grain is oblong truncate at both ends, reddish brown, with a prominent groove on the dorsal side; embryo occupying nearly half the length of the grain.

This grass grows abundantly in somewhat rich soils all over the Presidency and cattle eat it. It grows quickly and bears a fair amount of foliage.

Distribution.—Madras Presidency in the plains; also occurs in Ceylon.

[Pg 305]

Eragrostis pilosa, *Beauv.*

This is a densely tufted annual grass. Stems are usually erect, slender and simple, flaccid, 3 inches to 3 feet.

The *leaf-sheath* is compressed, glabrous and bearded with long hairs close to the mouth. The *ligule* is a ridge of hairs.

The *leaf-blade* is short, narrow, finely acuminate, 1-1/2 to 4 inches.

The *panicle* is oblong to pyramidal, flaccid, open or contracted erect or inclined, 2 to 8 inches; rachis is hairy or glabrous; branches are very fine filiform or capillary, more or less whorled, lower six inches long; branchlets are still finer and capillary.

Fig. 225.—Eragrostis pilosa.
1. A portion of a branch with spikelets; 2 and 3. empty glumes; 4. flowering glumes; 5. palea; 6. grain.

Spikelets are linear, grey tipped with purple, or often purplish, scattered, 1/8 to 1/5 by 1/30 to 1/20 inch, with pedicels shorter or longer than the spikelets. The *empty glumes* are hyaline, very unequal, nerveless or the second which is ovate-lanceolate and larger than the first faintly 1-nerved. The *flowering glumes* are ovate acute, paleate, 1/10 to 1/8 inch; palea is sub-persistent and keels of palea scaberulous. *Stamens* are three with small violet anthers. Grain is ellipsoid laterally pointed at the base.

This grass occurs in wet places or close to the margins of ponds, marshy situations all over the Presidency.

Distribution. — All over India and also in South Europe and most warm countries.

[Pg 306]

Eragrostis cynosuroides, *Beauv.*

This is a tall perennial grass freely branching from the base and with stout stolons covered with shining sheaths. The root-stock is stout and creeping. The stems are tufted, smooth, erect, with fascicles of leaves at the base 1 to 3 feet high.

The *leaf-sheath* is glabrous, slightly compressed, distinctly keeled, as broad or slightly broader than the blade at the mouth. *Ligule* is a line of short hairs.

The *leaf-blade* is linear, rigid, glabrous, acuminate with filiform tips, and finely serrulate margins, varying in length from 2 to 10 inches and the basal leaves sometimes reaching 20 inches.

The *panicle* is strict, erect, narrowly pyramidal, often interrupted, varying in length from 6 to 18 inches and breadth from 1/2 to 2 inches. Branches are many, short, crowded, densely clothed from the base with sessile, imbricating, much compressed deflexed spikelets.

Fig. 226.—Eragrostis cynosuroides.
1. A branch with spikelets; 2. flowering glumes with their palea; 3
and 4. empty glumes; 5 and 6. flowering glume and its palea.

The *spikelets* are secund, biseriate, shining, pale brown, 1/2 inch
long, up to 30-flowered. The *empty glumes* are unequal, the second
being the larger. The *flowering glumes* are coriaceous, ovate, acute as
long as the second or slightly longer, paleate, palea is sub-
coriaceous and shorter than the glume. *Stamens* are three. Grain is
obliquely ovoid, laterally compressed.

This grass grows usually in moist sandy loams, sand dunes, and
is very common on the Coromandel coast and in the Deccan Districts.

Distribution.—Throughout in the plains of India.

[Pg 307]

Eragrostis bifaria, *Wight Ex Steud.*

This is a densely tufted perennial grass. Stems are simple, erect, glabrous, somewhat compressed, 1 to 3 feet high, and the base clothed with the old remains of the leaf-sheaths.

The *leaf-sheath* is scaberulous, keeled. The *ligule* is a line of fine hairs.

The *leaf-blade* is wiry, narrow, linear, flexuous, rigid, acute, smooth, flat or complicate, keeled, 2 to 3 inches long and up to 1/6 inch wide.

The *spikes* are solitary, 10 to 12 inches long bearing spikelets uni-laterally.

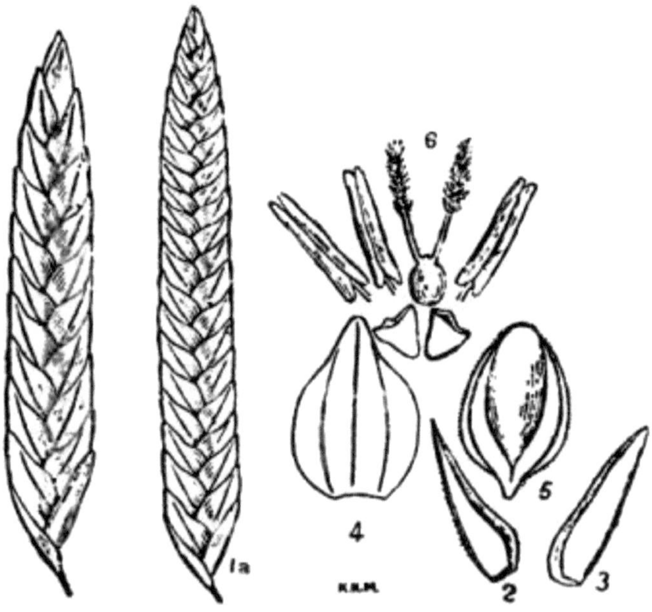

Fig. 227. — Eragrostis bifaria.
1 and 1a. Spikelets; 2. and 3. empty glumes; 4 and 5. the flowering glume and its palea; 6. the ovary, stamens and the lodicules.

The *spikelets* are ovate or ovoid to oblong, much compressed, usu-ally 15- to 20-flowered and up to 40 and then linear, 1/4 to 2/3 inch long, spreading, green or olive grey. The *empty glumes* are one-nerved and keeled. The *first glume* is longer than the second glume, very acute or acuminate. The *second glume* is smaller than the first, with stout rounded keel. The *flowering glumes* are as long or slightly shorter than the first glume, broadly ovate, sub-acuminate, with faint nerves and paleate; *palea* is shorter than its glume and with ciliate wings to the keel. *Stamens* are three. Grain is free.

This grass is very common in the plains in somewhat wet situations all over the Presidency.

Distribution.—Deccan Peninsula in India and also in Tropical Africa.

[Pg 308]

43. Oropetium, *Trin.*

A very small densely tufted erect annual. Leaves are filiform. The inflorescence is a simple slender curved spike. Spikelets are very minute, one-flowered, half immersed in the alternating distichous cavities of the rachis of the spike; rachilla is bearded. There are three glumes in the spikelet. The first glume is very minute, empty, hyaline and persistent. The second glume is linear-lanceolate, rigid, empty, persistent recurved when old, tip obtuse or emarginate. The third glume is shorter broader, hyaline, one-nerved, obtuse or truncate, *palea* is narrow with smooth keel. Lodicules are not found. Stamens are three. Grain is oblong terete and free.

[Pg 309]

Oropetium Thomæum, *Trin.*

This is a very small densely tufted annual grass, never exceeding 3 inches in height and with compressed slender, tough stems.

The *leaf-sheath* is compressed, membranous, short and open. The *ligule* is an erect lacerate membrane.

The *leaf-blade* is filiform, shorter or longer than the stem, erect or curved, coriaceous with the margins sparsely ciliate with long strict hairs, 1/2 to 1 inch long.

The *spikes* are solitary or fascicled curved on very short branches 1 to 1-1/2 inches long; rachis is green, undulating, tetragonous, with a broad central nerve on the flat faces.

Fig. 228.—Oropetium Thomæum.
1. Spike; 2. spikelet; 3 and 4. empty glumes; 5. flowering glume; 6 and 7. flowering glume and its palea; 8. the ovary, stamens and lodicules.

The *spikelets* are very small, one-flowered, half immersed in the alternating distichous cavities of the rachis. There are three *glumes* in the spikelet. The *first glume* is very minute, hyaline and sunk in the hollow of the rachis. The *second glume* is the longest, linear-lanceolate, rigid, tip obtuse or emarginate, slightly convex with a broad thickened centre and recurved in fruit. The *third glume* is shorter than the second, hyaline, broader obtuse, semi-circular in profile, excessively membranous, with the callus bearded and pale-ate; *palea* is smaller than the glume. There are three stamens. Grain is oblong, terete, free.

This small grass is very common all over the Presidency in the plains in moist places.

Distribution. — Plains of India, Burma and Ceylon.

[Pg 310]

[Pg 311]

GLOSSARY.

A

Acuminate
applied to the apex of a leaf having a gradually diminishing point, 49.

Acute
applied to the apex of a leaf distinctly and sharply pointed but not drawn out, 53.

Adventitious roots
roots which do not arise from the radicle or its subdivisions, but from parts other than these, 7.

Aleurone layer
a special peripheral layer in the grain of grasses, consisting of cells filled with proteid granules, 18.

Amplexicaul
applied to the base of the leaf when it embraces the stem, 12.

Apiculate
said of the apex when it has a sharp, short point.

Appressed
lying flat for the whole length of the part or organ, 59.

Articulate
jointed, 45.

Auricle
outgrowth at the sides close to the ligular region, 11.

Awned
having an *awn*, that is, a bristle-like appendage, especially on the glumes of grasses.

B

Bifarious
disposed in two rows or ranks on the two sides, 49.

Binate
in pairs, 53.

Blade
the expanded portion of a leaf, 2, 10.

Bristles
stiff hairs, 45.

Bulbous based
having an inflated base, 66.

Bulliform cells
thin walled cells occurring, at intervals, on the epidermis of some grasses, 35.

Bundle sheath
sclerenchymatous cells or fibres found round the vascular bundles of the monocotyledonous type, such as those of grasses, 20.

C

Callus
the projecting part or an extension of the flowering glume below its point of insertion, 168.

Caryopsis
a one-celled, one-seeded, superior fruit in which the pericarp has fused with the seed-coat.

Chartaceous
papery, i.e., thin and somewhat rough, 47.

Ciliate
fringed with hairs, 54.

Ciliolate
very sparsely fringed with hairs, 70.

Clavate
club-shaped, 104.

Clavellate
thickened towards the apex, 252.

Coleorhiza
the sheath of a monocotyledonous embryo which is pierced by the radicle during germination, 18.

Collar

the white or colourless band at the base of the blade of a grass leaf just where it joins the sheath, 3.

Conduplicate
folded together lengthwise, 12.

Convolute
rolled round from one margin to the other, so that one margin is inside and the other outside, 12.

Coriaceous
leathery, 49.

Corymbosely
arranged in corymbs, i.e., flat-topped flower clusters, 56.

Crinite
bearded with weak, long hairs, 137.

Crisped
curled, 59.

Cuneate
wedge-shaped or triangular, 49.

Cuspidate
tipped with a small triangular piece at the apex, 70.

D

Decumbent
reclining but with the upper part ascending, 80.

Digitate
lingered, arranged at the end of the stalk, 51.

Dioecious
having the sexes separated on two distinct individuals, 45.

Distichous
two-ranked or two-rowed, 19.
[Pg 312]

E

Embryo
young plant contained in the seed, 18.

Endodermis

the innermost layer of the cortex abutting on and forming a sheathing layer round the stele, 32.

Exodermis
the layer or layers of thickened cells beneath the piliferous layer of roots, 32.

Extra vaginal
applied to shoots or branches that come out piercing the leaf sheath in grasses, 9.

F

Fascicle
a cluster or bundle, 95.

Filiform
thread shaped, slender and thin, 54.

Flexuous
bent alternately in opposite directions, 62.

Foveolate
marked with small pits, 180.

G

Geniculately
bent abruptly so as to resemble a knee-joint, 118.

Geminate
in pairs, 59.

Germ-sheath
a sheath enclosing the bud or the plumule in a grain, 18.

Gibbous
convex or rounded, 77.

Glabrescent
slightly hairy but becoming glabrous, 89.

Glabrous
quite smooth without hairs, 89.

Glaucous
covered with a bloom, 160.

Glume

the chaffy two-ranked members found in the inflorescence of grasses.

H

Hirsute
covered with fairly long distinct hairs, 90.

Hyaline
colourless or translucent, 51.

I

Imbricate
overlapping, 49.

Internode
portion of a stem between two nodes, 2.

Intravaginal
growing out from inside the sheath.

Involucel
a ring of bracts surrounding several spikelets, 120.

K

Keeled
having a ridge along the length, 59.

L

Lemma
the flowering glume of a grass, 15.

Ligule
the thin, scarious projection found at the top of the leaf sheath where it joins the blade in grasses, 3.

Lodicule
a small scale outside the stamens in the flower of grasses.

M

Membranous
thin and semi-transparent, 51.

Monoecious
stamens and pistils on separate flowers, but on the same individual, 144.

Motor cells
large thin-walled cells occurring in the epidermis of the leaves of
some grasses, 35.

Mucronate
possessing a short and a straight point, 70.

N

Node
the part of the stem which has a leaf, or the knot in the grass stem, 2.
[Pg 313]

P

Palea
the inner glume in the spikelet of grasses, 4.

Pectinate
pinnatifid with narrow segments which are set close like the teeth of
a comb, 162.

Pericycle
the outermost zone of cells of the stele immediately within the endodermis, 32.

Phloëm
the portion of the vascular bundle towards the cortex, 19.

Pileole
another name for germ-sheath, or the sheath covering the plumule
in the grain, 18.

Piliferous
bearing hairs, 31.

Pistil
the female organ of a flower, consisting of the ovary, style and stig-
ma, 16.

Plumose
feathered, 51.

Prophyllum
the first scale-like leaf of a branch found where it joins the main
stem, 10.

Protandry (proterandry).

anthers ripening before the pistil in the same flower, 16.

Protogyny (proterogyny).
pistil ripening before the anthers in the same flower, 16.

Puberulous
slightly hairy, 62.

Pubescent
clothed with soft hair, 62.

Punctate
marked with dots, pits or glands, 63.

Pungent
ending in a rigid and sharp point, 59.

R

Raceme
a centrifugal or indeterminate inflorescence with stalked flowers,
13.

Rachilla
a secondary axis in the inflorescence of the grasses, the axis of the
spikelet, 13.

Rachis
axis of an inflorescence, 13.

Retuse
with a shallow notch at the apex, 67.

Rhizome
root-stock or under ground stem prostrate on the ground, 5.

Rugulose
somewhat wrinkled, 90.

S

Scaberulous
slightly rough due to the presence of short hairs, 69.

Scabrid
somewhat rough, 75.

Scale
a reduced leaf, 10.

Sclerenchyma
elongated cells with pointed ends and much thickened cell-wall.

Scutellum
the single cotyledon found in connexion with the embryo in grass grains, 18.

Secund
directed to one side only, 47.

Serrate
beset with small teeth on the margin, 83.

Setose
beset with bristles, 102.

Sheath
the tubular lower part of a leaf in grasses, 2.

Spathaceous
having a large bract enclosing a flower cluster, 104.

Spiciform
spike-like, 13.

Spike
an inflorescence with sessile flowers on an elongated axis, the older flowers being lower down and the younger towards the top, 13.

Squarrose
rough with outstanding processes, 120.

Stipe
a short stalk of a gynæcium, 90.

Stipitate
having a short stalk, 62.

Stolon
any basal branch which is disposed to root, 5.

Striolate
marked with very fine longitudinal parallel lines, 49.

Sub-coriaceous
somewhat leathery, 47.

Subulate
finely pointed, 121.

T

Triquetrous
three-sided or edged, 47.

Truncate
as if cut off at the end, 60.

Tumid
swollen, 66.

Turbinate
cone-shaped or top shaped, 120.

X

Xylem
the wood elements of the vascular bundle lying next to the phloëm, 19.

[Pg 314]

[Pg 315]